SECOND EDITION

NORTHEAST FERNS

A Field Guide to the Ferns and Fern Relatives
of the Northeastern United States

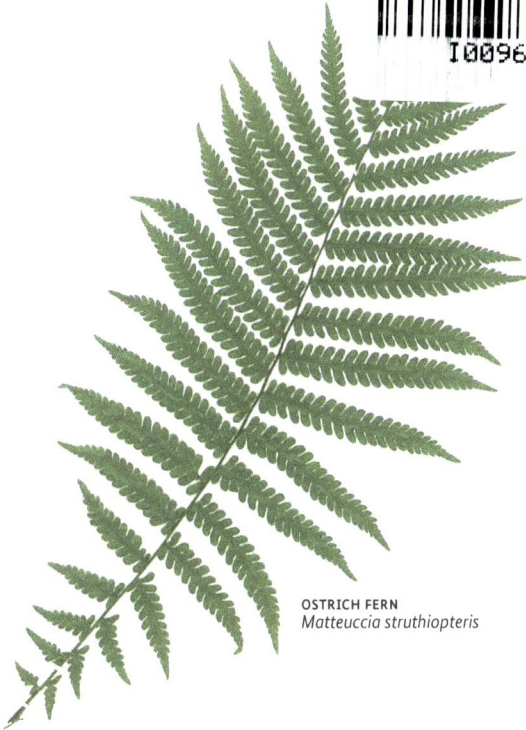

I0096417

OSTRICH FERN
Matteuccia struthiopteris

STEVE CHADDE

NORTHEAST FERNS
A Field Guide to the Ferns and Fern Relatives of the Northeastern United States
SECOND EDITION

Steve W. Chadde

ISBN: 978-1951682712

A Pathfinder Field Guide
Published by ORCHARD INNOVATIONS
Mountain View, Arkansas
Author email: *steve@orchardinnovations.com*

VERSION 1.0 01/2023

CONTENTS

PREFACE TO SECOND EDITION

THE SECOND EDITION OF NORTHEAST FERNS, now with full-color photographs, incorporates the fern and lycophyte classification proposed in 2016 by the Pteridophyte Phylogeny Group (PPG). This group of international scholars and fern specialists aimed to incorporate the latest research findings and the knowledge of those most familiar with ferns into a classification of the world's fern and lycophytes down to the genus level. Their work defined two classes (Lycopodiopsida, the lycophytes, and Polypodiopsida, the true ferns), 14 orders, 51 families, 337 genera, and an estimated 11,916 species.

In addition to nomenclatural changes, distribution maps have been updated to reflect changes since the publication of the first edition of *Northeast Ferns* in 2013.

Grateful acknowledgment is given to the Biota of North America Program (www.bonap.org) for permission to use their data to generate the distribution maps for the region. North American maps were adapted from those prepared for the Flora of North America project (floranorthamerica.org). Permission was provided by the New York State Museum for use of line drawings published in *Field Guide to Northeastern Ferns* (Bulletin 444) by Eugene C. Ogden (1981). Photographs were used under Creative Commons commercial use licenses, and I thank all those involved in documenting our ferns and other wild plants.

It is my hope that this new edition will continue to serve the needs of fern lovers of the region, and spark interest in the conservation of their habitats.

—STEVE CHADDE

RYAN HODNETT

Unfurling fronds of *Matteuccia struthiopteris*, OSTRICH FERN

INTRODUCTION

NORTHEAST FERNS IS A COMPREHENSIVE FIELD GUIDE to the ferns and fern relatives found in the states of Connecticut, Maine, Massachusetts, New Hampshire, New Jersey, New York, Pennsylvania, Rhode Island, and Vermont. Included are keys, descriptions, and distribution maps for 20 families and 38 genera of true ferns, and 3 families and 8 genera of lycophytes ('fern relatives': quillworts, clubmosses and spike-mosses). Note that in the following, the term 'fern' encompasses true ferns as well as lycophytes.

Ferns occur across the northeast in a wide variety of habitats, a region characterized by mountains and hills (part of the Appalachian Mountain chain), coniferous and deciduous forests, portions of the Atlantic Coastal Plain, and numerous lakes, rivers and wetlands. While most of our ferns prefer somewhat shaded locations in forests, a few species are adapted to growing in water or in saturated soils, while others survive in harsh alpine areas or on rock cliffs.

Background

Ferns are vascular plants that do not produce flowers or seeds, reproducing instead via the production of spores. Worldwide, ferns are the most abundant group of seedless vascular plants, with an estimated 9,200 to 12,000 species. The fossil record indicates that ferns originated during the Devonian period which began about 416 million years ago, and became abundant and with a diversity of forms during the Carboniferous Period (360–299 million years ago). During that period, giant ferns, horsetails resembling large bamboos, and clubmosses the size of trees were predominant in the warm, moist tropical forests of the lower latitudes. Sea level changes brought on by glacial cycles at the poles resulted in periodic inundation and subsequent burial of these forests. Over time, and with heat and pressure, the remains transformed into coal.

Toward the end of the Carboniferous Period, climatic changes led to a decline in many of the period's ferns. However, ferns again rose in importance in the Mesozoic Era (251–65 million year ago), and many of our modern species originated at that time. In the Cretaceous Period of the latter part of the Mesozoic Era (about 145 million years ago), flowering plants first appeared, gradually becoming predominant over the Earth's land surface.

Today, ferns occur in most terrestrial habitats and are also present in some aquatic situations. About 75% of fern species occur in the tropics. Ferns are also an important part of the understory vegetation in many forest communities and, with about one-third of the species growing on trunks and branches of trees, they are an important component of many epiphytic plant communities. Some species are very beneficial to humans, while some ferns such as the aquatic Kariba weed (*Salvinia molesta*) are troublesome, invasive weeds.

Ferns range in size from the very small, as in the mosquito ferns (*Azolla*) to tree-like, as in the tree ferns (tropical species not present in our region) which can grow to more than 20 meters tall and with leaves to 5 meters or more long.

7

Fern Reproduction

The fern life cycle usually involves an alternation of two free-living generations —sporophyte and gametophyte (illustration, page 9). Sporophytes are the typical, larger leafy form we most often see, and is the spore-producing stage of the life cycle. The gametophyte stage is the sexual phase of a fern. Both stages are photosynthetic. In contrast to nonvascular plants like mosses and liverworts, in ferns, the sporophyte generation is generally predominant and more developed. In seed plants, the gametophyte is no longer free-living but remains enclosed in the sporophyte.

Sporangia, masses of spore cases, are produced on the leaves (often termed fronds in ferns) of the sporophyte, or sometimes in cone-like strobili (as in the horsetails, *Equisetum*). In true ferns, the sporangia are most often on the underside of the leaf blade, and are often grouped into clusters called sori.

Within each sporangium, specialized cells undergo a series of mitotic (structural) divisions followed by meiosis (sexual division) that results in spores with half as many chromosomes as in the original sporophyte. More advanced ferns usually have 64 spores per sporangium, but more primitive ferns and fern relatives may have many more. When mature, the sporangium dries and opens, dispersing the spores into the air.

When a spore lands on a suitable surface, it germinates, the cells dividing and forming first a filament and then a heart-shaped gametophyte (or sometimes other shapes in some groups). Gametophytes resemble a moss, are small (usually less than 1 cm wide), and have a life span usually of only several months. They are the sexual phase of the life cycle in that they produce multicellular sex organs at maturity on the side of the gametophyte located away from light. Antheridia (male gametangia) are produced among the root-like rhizoids near the base of the plant. At maturity they open to release tailed spermatozoids.

Archegonia (female gametangia) are usually formed near the notch at the opposite end of the gametophyte. They are flask-shaped and contain a single egg cell. A droplet of water is needed for the spermatozoids to swim to an archegonium. When mature, the neck of the archegonium opens, and the spermatozoid swims down the opening to fuse with the egg. This forms a zygote with twice the number of chromosomes as the gametophyte. The zygote grows and develops into a new sporophyte, the gametophyte withers away, and the cycle is completed.

VARIATIONS

In some ferns and fern relatives, including members of the clubmosses (Lycopodiaceae) and moonworts and grape ferns (Ophioglossaceae), the gametophytes are not surface-dwelling and green. Instead, they are below the soil surface and nonphotosynthetic, often appearing as pale brown or yellow fuzzy cylinders or pads of tissue. These gametophytes are mycotrophic; that is, they receive their nutrients from soil fungi that establish connections with their rhizoids. Male and female organs are produced as in typical ferns, and the life cycle continues in the normal manner.

FERN LIFE CYCLE

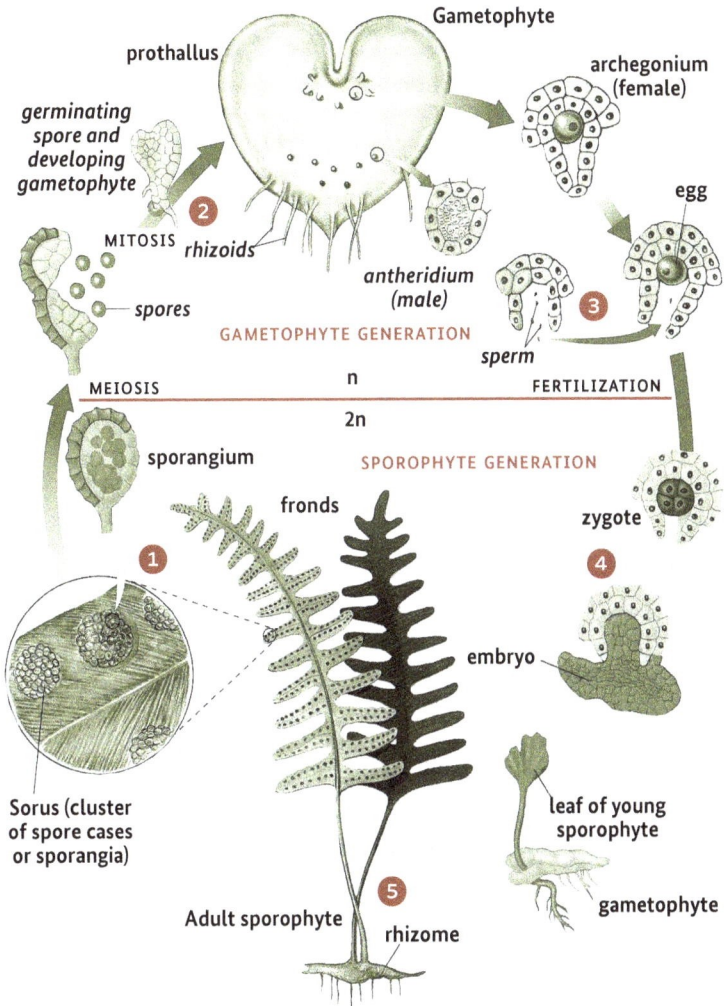

Gametophyte

prothallus

germinating
spore and
developing
gametophyte

archegonium
(female)

egg

2 MITOSIS rhizoids

spores

antheridium
(male)

3

GAMETOPHYTE GENERATION

sperm

MEIOSIS

n

FERTILIZATION

2n

sporangium

SPOROPHYTE GENERATION

zygote

fronds

1

4

embryo

Sorus (cluster
of spore cases
or sporangia)

leaf of young
sporophyte

5

gametophyte

Adult sporophyte rhizome

FERN LIFE CYCLE

1 Clusters (sori) of sporangia (spore cases) grow on the undersurface of mature fern fronds. **2** Released from its spore case, the haploid spore is carried to the ground, where it germinates into a tiny, usually heart-shaped, gametophyte, anchored to the ground by rhizoids (rootlike projections). **3** Under moist conditions, mature sperm are released from the antheridia and swim to the egg-producing archegonia that have formed on the underside of the gametophyte. **4** When fertilization occurs, a zygote forms and develops into an embryo within the archegonium. **5** The embryo eventually grows larger than the gametophyte and becomes a sporophyte.

Other ferns and fern relatives, including spike mosses (Selaginellaceae), quill-worts (Isoetaceae), and the aquatic ferns (Marsileaceae, Salviniaceae), depart from the typical life cycle in producing two different types of sporangia (heterospory). One sporangia produces numerous microscopic microspores that germinate to produce male gametophytes. The other form of sporangia produces many fewer and much larger megaspores (usually visible to the naked eye), which develop into female gametophytes.

VEGETATIVE REPRODUCTION

Many ferns display some form of vegetative reproduction. This may be simply the breaking of a rhizome into smaller pieces that become established as separate plants. This is common in horsetails (Equisetaceae), and if growing along a river, pieces of the rhizome may be widely dispersed during high-water periods.

Other ferns develop specialized structures for vegetative propagation. These include stolons (long, spreading stems that root at their tips and form new plants), bulbils on the leaves that can germinate to form new plantlets, and fronds that can produce roots where they come into contact with soil. A few species have specialized underground structures, such as tubers and similar offsets. Some species produce gemmae, which are specialized fragments of plants that break off and to form new plants. In a few species of filmy ferns (Hymenophyllaceae), and shoestring ferns (*Vittaria*), the ability to produce sporophytes has been lost and plants exist only as colonies of gametophytes spreading through the production of tiny air-dispersed gemmae.

APOGAMY

Apogamy is a form of asexual reproduction where the sporophyte grows directly from the gametophyte rather than from the fertilized egg. Apogamy is a widespread and important mechanism of reproduction, more common in ferns than in any other group of plants. Apogamous ferns are most common in environments with seasonal extremes of heat, cold, or drought. In the sporangia of these plants, a mechanism during the series of cell divisions results in spores with the same genetic makeup as the sporophyte plant. These 'diplospores' grow into gametophytes that produce new sporophytes directly from tissue near the notch of the gametophyte. Advantages of apogamy include faster development of the gametophyte and that standing water is no longer needed for fertilization to take place. Many apogamous ferns continue to produce antheridia with functional spermatozoids, which can be released and fertilize eggs on nearby gametophytes of related sexual species. Once formed, such hybrids are always apogamous and able to reproduce themselves.

Fern Structure

The key to identifying ferns is to know a few basic terms that describe the parts of the plant; these are discussed below and are used in the descriptions throughout the book (also see the **Glossary** beginning on page 293).

STEMS

Most ferns have specialized stems called *rhizomes* (synonym *rootstock*). Fern rhizomes are inconspicuous because they generally grow on or below the surface of the substrate (soil, moss or duff) on which the fern grows. Because the stems are in direct contact with the soil people often confuse stems with roots. Fern roots are generally thin and wiry in texture and grow along the stem. They absorb water and nutrients and help secure the fern to its substrate.

Rhizomes vary greatly in size, thickness and orientation. These modified stems are sometimes smooth but more often have scales or hairs, at least towards the growing tip. Stems can be short-creeping, with fronds that are somewhat scattered along the stem, such as in fragile fern (*Cystopteris*); stems can be long-creeping resulting in fronds scattered along the stem, as in bracken fern (*Pteridium*). Stems can also be vertical, producing circular clusters of leaves, as in most wood ferns (*Dryopteris*). In cross-section, most fern rhizomes appear as an irregular ring of vascular bundles. In some primitive ferns and fern relatives, the vascular system is a solid uninterrupted cylinder.

In some of the primitive ferns and in most fern relatives, other types of stems occur. Moonworts and grape ferns (Ophioglossaceae) usually have somewhat tuberous stems. Quillworts (Isoetaceae) have very short stout stems with the nodes very close together to form a swollen, cormlike base. Most clubmosses (Lycopodiaceae) have relatively unspecialized stems.

In horsetails (Equisetaceae), the underground stem is short to long-creeping, with wiry roots radiating from the underground stem's joints. The aboveground stems are hollow and vertically grooved or ridged. In most species, silica is incorporated into the stem's outside surface, giving plants a rough texture. Leaves are tiny and arise at the joints of the aboveground stem. Since the leaves are fused to each other around the stem, they form a sheath extending a short way up the stem from the joint. The color, number and tip shape of these sheathing leaves are good characters for distinguishing species of horsetails from each other. Some horsetails also have slender branches radiating from the joints. These branches give many *Equisetum* their 'horsetail' look.

Horsetails can have one kind of aboveground stem (monomorphic), or be dimorphic, with fertile and sterile aboveground stems. Spores are produced by cones situated at the tip of the aboveground stem.

FRONDS

The fern *frond* (synonym *leaf*) includes the entire aboveground leaf, including the stipe (synonyms petiole, stalk) and *leaf blade* (synonym *lamina*) and have a wide variety of forms. In most true ferns, the fronds are the dominant feature of the sporophyte and can be complex in their pattern of division. In most fern relatives and a few primitive ferns, however, the leaves are reduced and scale-like, needle-like or grass-like, with only a single vein.

In most ferns, the development of the leaf follows a pattern known as circinate vernation. This produces a characteristic *crozier* or *fiddlehead* as the leaf uncurls

PARTS OF A FERN

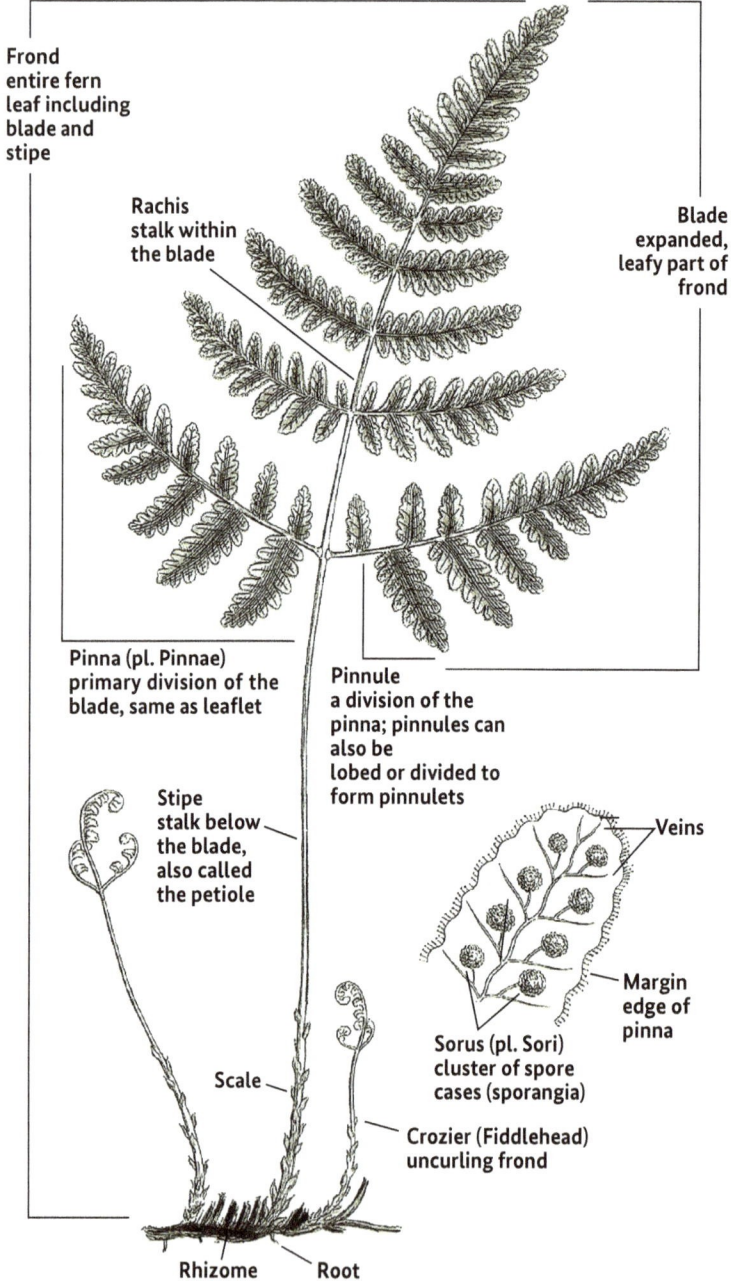

Frond
entire fern
leaf including
blade and
stipe

Rachis
stalk within
the blade

Blade
expanded,
leafy part of
frond

Pinna (pl. Pinnae)
primary division of the
blade, same as leaflet

Pinnule
a division of the
pinna; pinnules can
also be
lobed or divided to
form pinnulets

Stipe
stalk below
the blade,
also called
the petiole

Veins

Margin
edge of
pinna

Sorus (pl. Sori)
cluster of spore
cases (sporangia)

Scale

Crozier (Fiddlehead)
uncurling frond

Rhizome   **Root**

VEIN TYPES

netlike   not forked   once forked

twice or more forked   veins to margin   veins stopping before margin

and is the first structure to appear aboveground in spring (a crozier is the stylized staff of office carried by high-ranking church officials). The fern relatives and the moonworts and grape ferns (Ophioglossaceae) do not exhibit circinate vernation but expand by unfolding or by developing in an indefinite pattern.

The stipe of fern leaves may be circular, angled, or U-shaped in cross-section, and is sometimes hairy or scaly. There are one to several vascular strands or bundles of conductive tissue, and the number and shape of these in the stipe are often diagnostic for individual families or genera.

The leaf blade varies from *entire* (synonym *simple*) as in hart's-tongue fern (*Asplenium scolopendrium*) to **highly divided** as in lady fern (*Athyrium*). In true ferns, the most common type of division is is one or more times *pinnately compound*, less common are *pedate* and *palmate* patterns of division. In divided blades, the primary segments are called *pinnae* (synonym *leaflets*); a single segment is a *pinna*.

A pinna may be further divided into smaller segments (2-pinnate or bipinnate); these are called *pinnules* (synonym *subleaflets*) if fully divided from the pinna.

Pinnate means that the blade has distinct pinnae joined to the axis of the pinna by a narrow stalk. In many ferns, the pinnules are not cut to their midrib (only deeply lobed), and the term *pinnatifid* is used to describe this type of blade dissection.

FROND DIVISION

Simple
*undivided, not
compound*

Pinnatifid
blade cut partially to rachis

1-Pinnate
*blade divided into
pinnae*

2-Pinnate
*blade divided into
pinnae, the pinnae
divided into pinnules*

1-Pinnate Pinnatifid
*blade divided into
pinnae and the pinnae
lobed*

3-Pinnate
*blade divided into
pinnae, the pinnae
divided into pinnules,
the pinnules again
divided (pinnulets)*

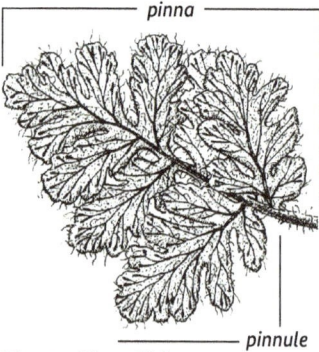

pinna

pinnule

2-Pinnate Pinnatifid
*blade divided into pinnae, the pinnae
divided into pinnules, and the pinnules
lobed*

Pedate
*blade palmately
divided, the segments
divided again in two*

14

SORI

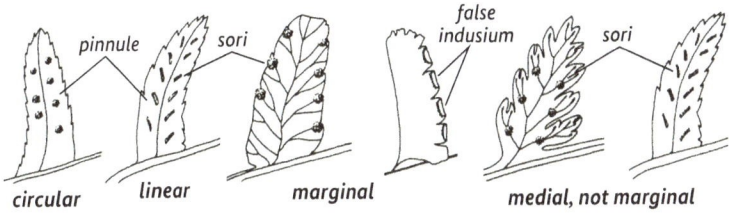

circular   linear   marginal   medial, not marginal

INDUSIA

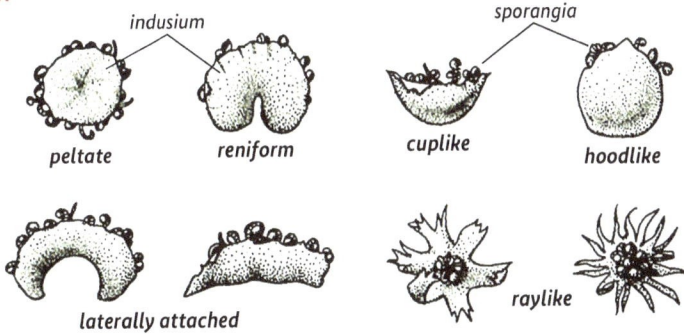

peltate   reniform   cuplike   hoodlike

laterally attached   raylike

ABOVE—FERN SORI AND INDUSIA

Sori (singular sorus) are groups of sporangia (singular *sporangium*), which contain spores. Sori are usually found on the underside of the blade. Young sori are often covered by a flap of protective tissue called the *indusium* (plural *indusia*). The shape and arrangement of the sori and indusia can be helpful characters for identifying unknown ferns. However, depending on the time of year, sori and indusia may be too immature or may have already been shed, and are therefore of little use for identification.

LEFT—FERN LEAF BLADE DIVISION

- 1-PINNATE leaf blades divided into pinnae, the pinnae more or less entire and not deeply lobed.
- 1-PINNATE-PINNATIFID leaf blades divided into pinnae, and the pinnae deeply lobed but not completely divided to midrib.
- 2-PINNATE leaf blades divided into pinnae, and the pinnae divided into pinnules.
- 2-PINNATE-PINNATIFID leaf blades divided into pinnae, and the pinnae divided into pinnules, but the pinnules deeply lobed and not completely divided to the midrib.
- 3-PINNATE leaf blades thrice pinnately divided.
- 3-PINNATE-PINNATIFID leaf blades thrice pinnately divided, and the pinnulets (smallest segments) deeply lobed but not completely divided to midrib.

SORUS OF MALE FERN

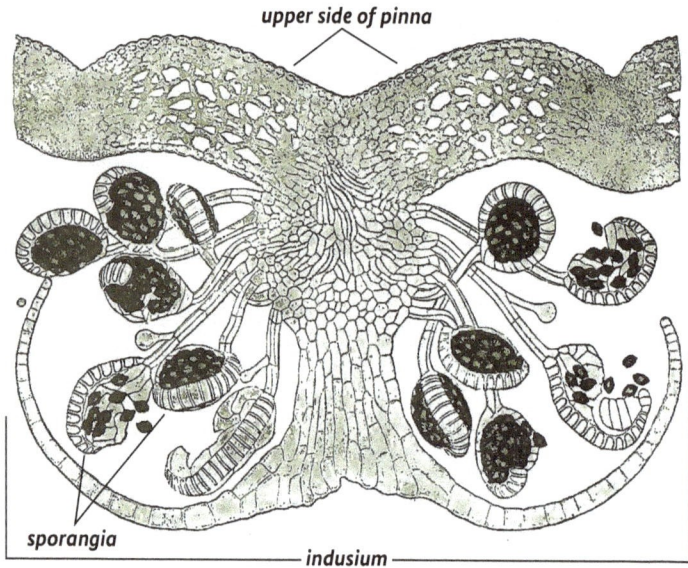

upper side of pinna

sporangia

indusium

CROSS-SECTION through single sorus of male fern (*Dryopteris filix-mas*). The indusium is shaped like an inverted open umbrella and covers the stalked sporangia which are opening to release their spores. Circled area at right outlines a single sorus on the pinna underside.

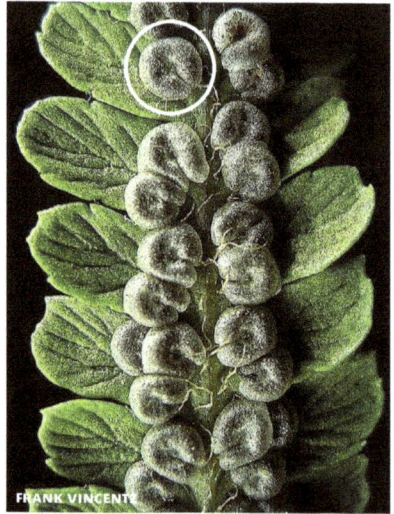

FRANK VINCENTI

The continuation of the stipe as the central axis of the leaf blade is known as the *rachis*. Pinnae are attached to the rachis. The midrib of the pinna and the pinnule is termed the *costa* (plural *costae*). The smaller veins of the pinnule may be unbranched or branched, and free (not rejoining) to variously anastomosed (joined to form a netlike pattern). Most of our ferns have free veins, but several genera have netlike venation as in the chain ferns (*Anchistea, Lorinseria*).

Fern fronds may be smooth or variously covered with hairs, scales, or both. In some species, the leaves are glandular and sticky. Other species secrete a powdery farina, usually on the leaf underside, which may be white or bright yellow or orange. Fronds also vary greatly in thickness and texture. The thinnest leaves occur in the filmy ferns (Hymenophyllaceae), with leaves often only 1–2 cell layers thick. In contrast, other species have thick leathery leaves, or leaves with dense

REPRESENTATIVE FERN SPORES

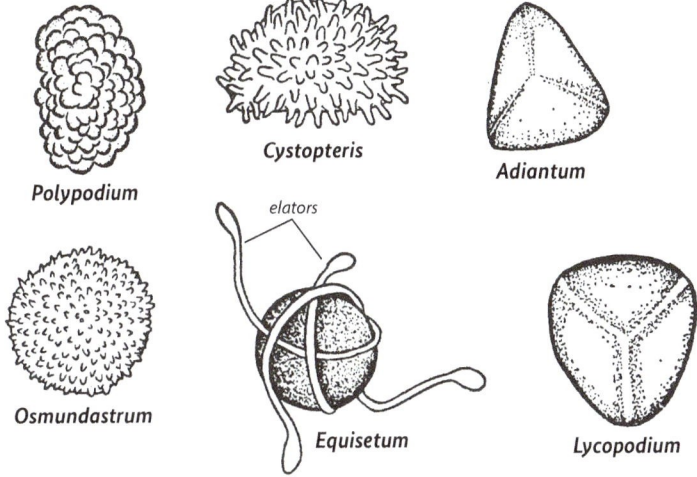

Cystopteris

Polypodium

Adiantum

elators

Osmundastrum

Equisetum

Lycopodium

coverings of hairs, scales, glands, or farina. These are believed to be adaptations to dry, harsh habitats such as those found on exposed rock cliffs.

SPORANGIA

In most ferns, the *sporangia* (singular *sporangium*) are spore-cases produced on the underside of normal fronds. In fern relatives and a few primitive ferns, relatively large sporangia are borne in either cone-like strobili at the tips of stems or branches, or in the axil at the leaf base. Some ferns have **dimorphic fronds,** with sterile or vegetative fronds (the *trophophyll*), and fertile fronds (*sporophyll*) of different size and shape. In other ferns, the frond is divided into specialized fertile and sterile portions.

The position, shape, and other details of the sporangia on the leaf underside are often characteristic of particular families or genera and are an important factor in fern classification. In most ferns, the sporangia are grouped into discrete clusters known as **sori** (singular *sorus*). Sori may be round to linear, positioned along the margin or towards the midvein (costa), on the blade surface or in a groove or channel. In some species, sporangia may entirely cover the blade underside (a condition termed *acrostichoid*). In some primitive ferns, sporangia are only sparsely scattered along a few veins.

In some species, such as maidenhair fern (*Adiantum*), the developing sori are protected by a recurved leaf margin termed a **false indusium**. In other ferns, a covering of deciduous scales, or a more permanent small flap of tissue, the indusium is present. Indusia vary in shape, size, texture and persistence, ranging from umbrella-shaped to round to linear. In the water ferns (Marsileaceae, Salviniaceae), sporangia are contained within hard capsules called *sporocarps*.

The sporangia themselves are usually positioned on a thickened vein tip or along a portion of a vein. In most ferns, the sporangium consists of a stalk and a

17

capsule. The capsule may have an annulus, a ring or region of cells with only some of the walls thickened; the annulus acts somewhat like a spring to project the spores out of the sporangia.

SPORES

Spores are the main means by which ferns are dispersed to form new populations. In most ferns, spores are relatively long-lived and impervious, and fern spores have been successfully germinated after more than a hundred years of storage. After the sporangia opens to shed the mature spores, most fall within a few meters of the parent. However, spores have been recovered from the stratosphere during high-altitude atmospheric sampling, and ferns are among the most successful colonists of isolated ocean islands, testimony to their ability to successfully travel very long distances. In a few ferns, the spores are shorter-lived, relatively thin-walled, green and photosynthetically active, allowing for rapid establishment of new plants following dispersal.

Developmentally, spores are the direct products of meiosis, which begins with a single spore mother cell undergoing two separate rounds of division, and yields a tetrad that breaks into four individual spores. Two main spore types are recognized: trilete (tetrahedral) spores vary from nearly spherical to somewhat three-angled and have a three-branched scar where each spore was attached to the others in the tetrad. Monolete (bilateral) spores are ellipsoid to bean-shaped and have a linear attachment scar along one side. The attachment scar is usually where the spore ruptures during germination.

Mature spores vary greatly in size and surface sculpturing (see figure on page 17). Small spores of some species are only about 15 microns in diameter, whereas megaspores of some *Selaginella* species may approach 1 mm. The surface sculpturing is often diagnostic for various species, genera or families (as in the quillworts, Isoetes), ranging from smooth to wrinkled, spiny, or with wing-like ridges.

GAMETOPHYTES

Upon germination of the spore, cell division produces first a threadlike structure. In most ferns, this continues to divide and eventually develops into the mature gametophyte. The typical fern gametophyte is flat and heart-shaped, with a notch and 2 lobes at one end and with the other end narrowed or rounded. Gametophytes vary in size from a few millimeters to about 1 cm wide. Other shapes occur but are not common; these include gametophytes that are filamentous, strap-shaped, or irregularly lobed. In ferns with underground mycorrhizal gametophytes, these mature mostly to either tubular or cushion-shaped structures.

Similar to mosses, gametophytes lack vascular tissue and roots. However, hair-like structures called rhizoids absorb water and nutrients and help anchor the gametophyte to the substrate. The gametangia (male and female sex organs) generally are formed on the side of the gametophyte away from the light (except in subterranean gametophytes). The male organs (antheridia) are positioned among the rhizoids, are spherical, and consist of a covering of cells enclosing the spermatozoids. The female organs (archegonia) are usually located near the notch on a slightly thickened pad of tissue. They are flask-shaped and somewhat sunken into the tissue. The neck of the archegonium consists of four columns of cells that separate at maturity, opening a canal and exposing the egg cell at the base for fertilization by the spermatozoid.

Fern Classification

At the genus and species level, fern names have remained fairly stable in recent years. Higher level classifications of ferns, however, remain somewhat controversial, as relationships between large groups continue to be studied and clarified. For our purposes, an emphasis is placed on family divisions and the fern families are arranged alphabetically in the book, first by the true ferns followed by the lycophytes or fern relatives. See page 20 for the higher level (subclass, order) arrangement of northeastern ferns and fern relatives.

In the past, ferns had been loosely grouped with other spore-bearing vascular plants, these often called "fern allies" or "lycophytes." However, recent studies suggest an important dichotomy within vascular plants, separating the fern relatives or lycophytes (less than 1% of all vascular plant species) from a group termed the euphyllophytes. Euphyllophytes comprise two major groups: the spermatophytes (seed plants), which number more than 260,000 species, and the monilophytes (ferns), with over 11,900 species worldwide, including horsetails, whisk ferns, and all eusporangiate and leptosporangiate ferns.

Genetic studies also reveal surprises about the relationships among true ferns and fern allies. True ferns appear to be closely related to horsetails, and in fact these plants are now grouped within the true ferns. Also, plants commonly called fern relatives (clubmosses, spike-mosses and quillworts) are not closely related to the true ferns. However, in the tradition of past fern guides, both true ferns and fern relatives are included so that all spore-bearing vascular plants in the region can be identified.

Economic Importance

Only a handful of fern species are economically important. Perhaps the best known use is horticultural, and many species are used in landscaping, as houseplants, and by florists. Pieces of the root mantles covering stems of tree ferns ('orchid bark') are used as a substrate for growing orchids and epiphytic plants.

An interesting phenomenon called *pteridomania* ("fern craze") occurred in Great Britain in the Victorian period of the mid- to late-1800s. Anything fern-related was highly popular, including illustrated fern books, growing ferns in gardens and greenhouses, and objects and art with fern motifs.

Ferns have been used in handicrafts by Native Americans. For example, the twining stipe of some members of the Climbing Fern Family (Lygodiaceae) and the lustrous stipes of maidenhair (*Adiantum*), were used in basket-making and in bracelets. Leaves of bracken fern (*Pteridium*) have been used to make a green dye.

Clubmosses (Lycopodiaceae) have an interesting history of uses. The spores

FAMILY RELATIONSHIPS

The Pteridophyte Phylogeny Group (PPG, 2016). defined two classes (**Polypodiopsida**, the true ferns, and **Lycopodiopsida**, the lycophytes), subdivided (worldwide) into 14 orders, 51 families, 337 genera, and an estimated 11,916 species. Present in the Northeast region are the following:

True Ferns

CLASS POLYPODIOPSIDA

SUBCLASS EQUISETIDAE

ORDER EQUISETALES
- FAMILY Equisetaceae

SUBCLASS OPHIOGLOSSIDAE

ORDER OPHIOGLOSSALES
- FAMILY Ophioglossaceae

SUBCLASS MARATIIDAE

ORDER HYMENOPHYLLALES
- FAMILY Hymenophyllaceae

ORDER OSMUNDALES
- FAMILY Osmundaceae

ORDER POLYPODIALES
- FAMILY Aspleniaceae
- FAMILY Athyriaceae
- FAMILY Blechnaceae
- FAMILY Cystopteridaceae
- FAMILY Dennstaedtiaceae
- FAMILY Diplaziopsidaceae
- FAMILY Dryopteridaceae
- FAMILY Pteridaceae
- FAMILY Onocleaceae
- FAMILY Polypodiaceae
- FAMILY Thelypteridaceae
- FAMILY Woodsiaceae

ORDER SALVINIALES
- FAMILY Marsileaceae
- FAMILY Salviniaceae

ORDER SCHIZAEALES
- FAMILY Lygodiaceae
- FAMILY Schizaeaceae

Lycophytes

CLASS LYCOPODIOPSIDA

ORDER ISOETALES
- FAMILY Isoetaceae

ORDER LYCOPODIALES
- FAMILY Lycopodiaceae

ORDER SELAGINELLALES
- FAMILY Selaginellaceae

contain nonvolatile oils that made them useful as a dry lubricant. They have also been used to keep latex products like surgical gloves from sticking together (now mostly discontinued since the spores can be a skin irritant to some people). Other uses of the spores have been in flash powder for photography and in fingerprint powder in forensic investigations.

Various ferns are eaten as food, with the young fronds prepared like a vegetable. The most important edible species is ostrich fern (*Matteuccia struthiopteris*), whose fiddleheads are available in many markets of the region in late spring. In parts of Asia, *Diplazium esculentum* is grown for food. Formerly, fiddleheads of bracken fern (*Pteridium aquilinum*) were eaten, but studies have linked this species to stomach cancer, and it should be avoided.

The most economically valuable species of fern is mosquito fern (*Azolla*), a tiny floating aquatic fern. For many centuries, farmers in parts of eastern Asia have inoculated rice paddies with this plant to increase the yield of their harvest. The hollow chambers in *Azolla* leaves contain symbiotic cyanobacteria (*Anabaena azollae*) that convert atmospheric nitrogen into the nitrate form, an important plant nutrient. Research is ongoing to identify superior strains of this fern.

Several ferns have negative economic impacts because of their weediness. The most notable are species of *Salvinia* and *Pteridium*. Kariba weed (or giant salvinia, *Salvinia molesta*) is a floating aquatic fern that is aggressively weedy throughout warmer parts of the world. Native to Brazil, it is now present in California, Texas, and the southeastern states. Kariba weed can form a surface mat several inches thick, preventing light and oxygen from penetrating into the water, choking out other aquatic plants as well as devastating fisheries. Bracken fern (*Pteridium aquilinum*) is widely distributed, and has an extensive, deep, creeping rhizome sometimes several hundred meters in length. Bracken fern quickly invades open habitats, out-competing other plants. Plants are toxic to livestock, rendering large areas unusable, especially in parts of Europe.

Fern Conservation

A number of our ferns are rare due to uncommon habitats or to small population sizes. Some of our species are afforded legal protection under state law (endangered or threatened); in the species descriptions, the status is noted under the conservation status heading. In general, when 'botanizing,' disturb habitats as little as possible, and refrain from collecting plants; in most cases, photographs will serve as adequate documentation.

The following terms are used to describe rarity within a state. Endangered and threatened status provides legal protection to a species:

• *Endangered*: a species threatened with "extirpation" (extinction within a single state or area and not across its entire range).

• *Threatened*: a species whose survival within a state is not in immediate danger, but for which a threat exists; continued stress on the species may result in its becoming endangered.

Each state maintains lists of species of conservation concern or has a natural heritage program to monitor populations of rare species; more information is available at the websites below.

NATURAL HERITAGE INFORMATION

- **Connecticut** www.ct.gov/deep
- **Maine** www.maine.gov/doc/nrimc/mnap/features/rareplant.htm
- **Massachusetts** www.mass.gov/eea/agencies/dfg/dfw/natural-heritage/
- **New Hampshire** www.nhdfl.org
- **New Jersey** www.nj.gov/dep/parksandforests/natural/heritage/
- **New York** https://www.dec.ny.gov/23.html
- **Pennsylvania** www.naturalheritage.state.pa.us
- **Rhode Island** https://dem.ri.gov
- **Vermont** https://anr.vermont.gov/

ABBREVIATIONS

CT	Connecticut	cm	centimeter
ME	Maine	dm	decimeter
MA	Massachusetts	mm	millimeter
NH	New Hampshire	m	meter
NJ	New Jersey	ca.	circa, approximately
NY	New York	n	north
PA	Pennsylvania	s	south
RI	Rhode Island	e	east
VT	Vermont	w	west

ILLUSTRATED KEY TO FAMILIES AND GENERA

1 Plant a free-living gametophyte, resembling a thallose liverwort; rare in dark, moist recesses in rock overhangs in NY and PA . . . . PTERIDACEAE, page 274
. . . . *Vittaria appalachiana*, page 296

1 Plant a sporophyte, consisting of a stem and well-developed leaves more than 1 cell thick (except in *Crepidomanes*), generally reproducing by spores 2

2 Plants aquatic, either floating and unattached, or rooting and often completely submersed
. . . . KEY A, page 22

2 Plants of various wetland, upland, and rock habitats 3

3 Leaves not "fern-like;" unlobed, variously awl-shaped, scalelike, or grasslike (the lycophytes will key here plus two species of *Asplenium*) . . . . KEY B, page 24

3 Leaves "fern-like," variously lobed or divided; true ferns will key here (the sub-keys are based on size of frond and habitat, either soil or rock). 4

4 Plants small, leaf blades (not including the stipe) small, less than 30 cm long or wide (some species will key both here and in the next lead) 5

4 Plants larger, leaf blades medium to large, more than 30 cm long or wide 6

5 Plants growing on rock, rock walls, or over rock in thin soil, or rarely on the bark of trees
. . . . KEY C, page 27

5 Plants terrestrial, growing in soil, not associated with rock outcrops . . . . . KEY D, page 31

6 Plants growing on rock, or over rock in thin soil mats or pockets of soil . . KEY E, page 34

6 Plants growing in soil, not associated with rock outcrops . . . . KEY F, page 38

KEY A — ferns growing in water, floating or rooted and submersed

1 Plants aquatic, floating . . . . SALVINIACEAE, page 221
. . . . *Azolla*, page 221

1 Plants aquatic, rooted, sometimes exposed on shores as water levels drop 2

2 Plants resemble a clover, with 4 terminal leaf segments . . . . . MARSILEACEAE, page 157
. . . . *Marsilea*, page 157

2 Leaves linear, from a swollen, corm-like base. . . . . ISOETACEAE, page 246
. . . . *Isoetes*, p. 246

SALVINIACEAE
Azolla

MARSILEACEAE
Marsilea

ISOETACEAE
Isoetes

KEY B — leaves unlobed, awl-shaped, scalelike, or grasslike, and not "fern-like"

1   Stems obviously jointed; leaves small and scalelike, in a whorl from the joints or nodes, or sometimes absent; spores borne in a terminal conelike strobilus covered with peltate scales (i.e., scales more or less round and attached at middle like an umbrella)............ EQUISETACEAE, page 132
.... *Equisetum*, page 132

1   Stems not jointed; leaves scalelike or larger, but if scalelike not in whorls from the nodes; spores borne variously, but if in a terminal strobilus the scales not peltate........... 2

2   Leaves linear, grasslike, 1-50 cm long.................. 3

2   Leaves various (scalelike, awl-like, moss-like, or flat), but not linear and grasslike 4

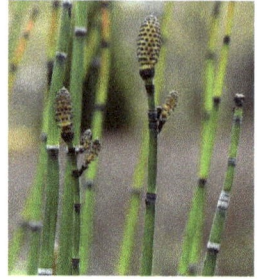

EQUISETACEAE
Equisetum

3   Leaves straight and stiff, arching, or flaccid, from a 2-3-lobed corm; sporangia borne in the expanded leaf bases..
.... ISOETACEAE, page 246
.... *Isoetes*, page 246

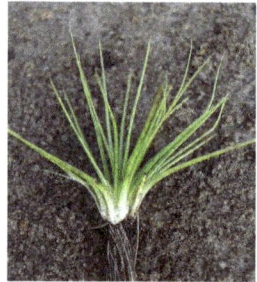

ISOETACEAE
Isoetes

3   Leaves notably spiral-curly, from a short-creeping rhizome; sporangia borne in 2 rows at the expanded tip of the fertile leaves..................... SCHIZAEACEAE, page 224
.... *Schizaea*, page 224

4   Leaves very numerous and overlapping along creeping, ascending, or erect stems; the leaves usually scalelike or awl-like, 0.5-2 (-3) mm wide, typically sharp- or hair-tipped; sporangia borne in strobili........................... 5

4   Leaves not as above 11

SCHIZAEACEAE
Schizaea

5   Sporangia borne in flattened or 4-sided strobili sessile at tips of leafy branches; spores and sporangia of two sizes, the megasporangia larger and borne at base strobili......
.... SELAGINELLACEAE, page 284
.... *Selaginella*, page 284

5   Sporangia borne either in axils of normal foliage leaves, or in strobili sessile at tips of leafy branches, or stalked on specialized branches with fewer and smaller leaves; spores and sporangia of one size (Lycopodiaceae) 6

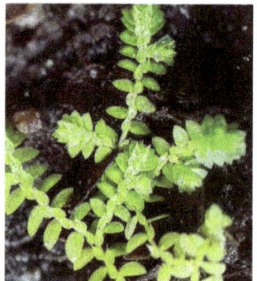

SELAGINELLACEAE
Selaginella

NOTE: See species page for photographer credit.

24

6 Leafy stems erect, simple or branched, the ultimate branches upright; sporophylls like the sterile leaves or only slightly smaller, in annual bands along the stem; vegetative reproduction by leafy gemmae near tip of stem . . . .
. . . . LYCOPODIACEAE, page 252
. . . . *Huperzia*, page 268

6 Leafy stems prostrate or erect, if erect then generally branched, the ultimate branches spreading (horizontal) or ascending; sporophylls differing from sterile leaves, either broader and shorter, or more spreading, grouped into terminal cones; lacking vegetative reproduction by gemmae .
. . . . 7

7 Plants of wetlands, mostly on moist or wet sand or peat; leaves herbaceous, pale or yellow-green, dull, deciduous; leafy stems creeping; rhizome dying back annually to an underground vegetative tuber at tip. . . . .
. . . . LYCOPODIACEAE, page 252
. . . . *Lycopodiella*, page 274

7 Plants of uplands, mostly in moist to dry soils; leaves stiff, bright to dark green, shiny, evergreen; leafy stems mainly erect, treelike, fanlike, or creeping (if creeping, then the leaves with a hairlike tip); rhizome trailing, perennial 8

8 Shoots flat-branched, 1-5 mm wide (including the leaves); leaves scalelike, dimorphic, overlapping and appressed to stem, in 4 ranks; strobili on long, branched stalks . . . .
. . . . LYCOPODIACEAE, page 252
. . . . *Diphasiastrum*, page 259

8 Shoots round-branched, usually 5-8 mm wide (including the leaves), leaves awl-shaped, monomorphic (though sometimes differing in size), separate, spreading or ascending, in 6 ranks; strobili sessile at stem tips 9

9 Strobili on elongate, sparsely leafy stalks at tips of leafy upright branches; leaves with hairlike tip . . . .
. . . . LYCOPODIACEAE, page 252
. . . . *Lycopodium*, page 278

9 Strobili not stalked, borne immediately above densely leafy portion of upright branches; leaves tapered to a narrow tip
. . . . 10

LYCOPODIACEAE
Huperzia

LYCOPODIACEAE
Lycopodiella

LYCOPODIACEAE
Diphasiastrum

LYCOPODIACEAE
Lycopodium

10 Upright leafy stems 10 mm or more wide (including the leaves), branched 1-4 times; leaves tipped with a small sharp spine; horizontal shoots creeping on ground surface
. . . . **LYCOPODIACEAE**, page 252
. . . . *Spinulum*, page 281

10 Upright leafy stems 3-8 mm wide (including the leaves), treelike or fanlike; leaves tapered to tip, spine absent; horizontal shoots underground **LYCOPODIACEAE**, page 252
. . . . *Dendrolycopodium*, page 254

11 Plants with 1 (-several) leaves, the sterile leaf blade ovate, entire-margined, obtuse, the longer fertile portion with 2 rows of sporangia somewhat embedded in it . . . .
. . . . **OPHIOGLOSSACEAE**, page 164
. . . . *Ophioglossum*, page 181

11 Plant with many leaves, generally 5 or more, not divided into separate sterile and fertile segments, the leaves strap-shaped, or lance-shaped and with a long-tapering tip (this often with a plantlet which can root to form new plants) .
. . . . **ASPLENIACEAE**, page 43
. . . . *Asplenium rhizophyllum*, page 54
. . . . *Asplenium scolopendrium*, page 58

LYCOPODIACEAE
Spinulum

LYCOPODIACEAE
Dendrolycopodium

OPHIOGLOSSACEAE
Ophioglossum

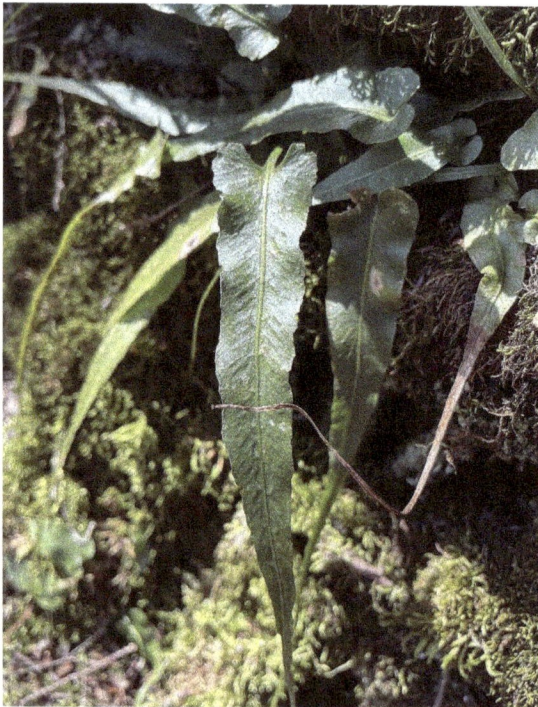

ASPLENIACEAE
Asplenium rhizophyllum

ASPLENIACEAE
Asplenium scolopendrium

KEY C — small ferns, growing on rock, on rock in thin soil, or (rarely) on trees

1 Fronds pinnatifid or 2-pinnatifid, most of the pinnae not fully divided from one another (the rachis winged by leaf tissue for most or all its length) 2

1 Fronds pinnate, pinnate-pinnatifid, 2-pinnate, or even more divided (rachis naked for most of its length, but often winged in upper portion) 5

2 Fronds 2-pinnatifid, at least the lowermost pinnae deeply lobed. 3

2 Fronds 1-pinnatifid 4

3 Lowermost (and other) pinnae with numerous, nearly even lobes; native ferns . . . . . . THELYPTERIDACEAE, page 225

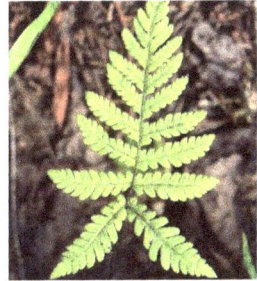

THELYPTERIDACEAE
Phegopteris

3 Lowermost pinnae with a few, irregular lobes (the upper pinnae unlobed); introduced fern, reported from NY . . . . PTERIDACEAE, page 207
. . . . *Pteris multifida*, page 207

PTERIDACEAE
Pteris multifida

ASPLENIACEAE
Asplenium platyneuron

4 Blades with long, narrow tapering tip, upper portion of blade unlobed or only slightly lobed; sori elongate . . . .
. . . . ASPLENIACEAE, page 43
. . . . *Asplenium* (in part), page 43

ASPLENIACEAE
Asplenium

POLYPODIACEAE
Polypodium virginianum

4 Fronds without a long, narrow tapering tip; blade lobed for most of its length; sori round . . . .
. . . . POLYPODIACEAE, page 202
. . . . *Polypodium*, page 202

POLYPODIACEAE
Polypodium

5 Fronds 1-pinnate or 1-pinnate-pinnatifid 6

5 Fronds 2-pinnate or more divided. 10

6 Fronds very delicate and algae-like; plants of caves and rock hollows . . . . HYMENOPHYLLACEAE, page 152
. . . . *Crepidomanes*, page 152

6 Fronds thicker, herbaceous, to leathery in texture, more than 1 cell thick; sori otherwise; plants of various habitats
. . . . 7

7 Pinnae more than 1 cm wide; fronds leathery; veins rejoining to form a netlike pattern; introduced, reported from MA and NJ . . . . DRYOPTERIDACEAE, page 102
. . . . *Cyrtomium falcatum*, page 102

7 Pinnae less than 1 cm wide; fronds herbaceous to somewhat leathery; veins free, not rejoining and netlike 8

8 Sori on the undersurface of the leaf, away from the margins
. . . . ASPLENIACEAE, page 43
. . . . *Asplenium*, page 43

8 Sori on the undersurface of the leaf, along margins and more-or-less hidden beneath either the unmodified inrolled leaf margin or under a modified, reflexed false indusium. 9

9 Stipes green to straw-colored for at least the upper 1/3, rachis green; fronds dimorphic, the fertile longer than the sterile and with narrower segments. . . . .
. . . . PTERIDACEAE, page 207
. . . . *Cryptogramma*, page 212

9 Stipes and rachis dark brown to almost black throughout; fronds similar or somewhat different . . . .
. . . . PTERIDACEAE, page 207
. . . . *Pellaea*, page 216

HYMENOPHYLLACEAE
Crepidomanes

DRYOPTERIDACEAE
Cyrtomium falcatum

ASPLENIACEAE
Asplenium

PTERIDACEAE
Cryptogramma

PTERIDACEAE *Pellaea*

10 Blade broadly triangular in outline. . . . .
. . . . **CYSTOPTERIDACEAE**, page 78
. . . . *Gymnocarpium*, page 92

10 Blade elongate, mostly lance-shaped, generally 4x or more as long as wide, not notably triangular in outline. **11**

CYSTOPTERIDACEAE
Gymnocarpium

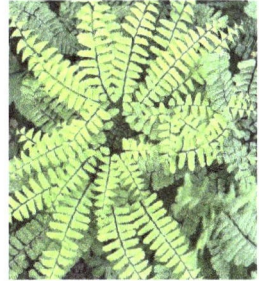

11 Sori on margins, usually more-or-less hidden under the in-rolled margin of the pinnule. **12**

11 Sori not on margins, either naked, or slightly to strongly hidden by indusia **14**

12 Sori round or oblong, distinct and separate along the pinnule margins; fronds bright-green, smooth, herbaceous, delicate, and flexible . . . . **PTERIDACEAE**, page 207
. . . . *Adiantum*, page 208

12 Sori continuous along the pinnule margins; fronds mostly dark-green, often hairy, leathery, tough, and stiff **13**

PTERIDACEAE
Adiantum

PTERIDACEAE
Myriopteris lanosa

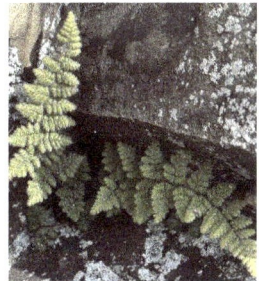

13 Fronds 2-3-pinnate, more or less densely hairy . . . .
. . . . **PTERIDACEAE**, page 207
. . . . *Myriopteris*, page 214

PTERIDACEAE
Myriopteris

PTERIDACEAE
Pellaea atropurpurea

13 Fronds 1-2-pinnate; smooth, or sparsely and inconspicuously hairy . . . . **PTERIDACEAE**, page 207
. . . . *Pellaea*, page 216

PTERIDACEAE
Pellaea

29

14 Blades 3-12 cm long; sori elongate, covered by a flap-like, entire indusium . . . . **ASPLENIACEAE**, page 43

. . . . *Asplenium*, page 43

14 Blades 4-30 (-50) cm long; sori round, surrounded or covered by an entire, fringed, or divided indusium **15**

15 Veins reaching margin; indusium attached under one side of sorus, hoodlike or pocketlike, arching over sorus; stipes smooth or sparsely covered with scales, stipe bases not persistent . . . . **CYSTOPTERIDACEAE**, page 78

. . . . *Cystopteris*, page 78

15 Veins ending short of margin; indusium attached under sorus, cuplike (divided into 3-6 lobes which surround the sorus from below) or of numerous tiny hairs, which extend out from under sorus on all sides; stipes often densely covered with scales, stipe bases persistent. . . . .

. . . . **WOODSIACEAE**, page 236

. . . . *Woodsia*, page 236

ASPLENIACEAE
Asplenium

CYSTOPTERIDACEAE
Cystopteris

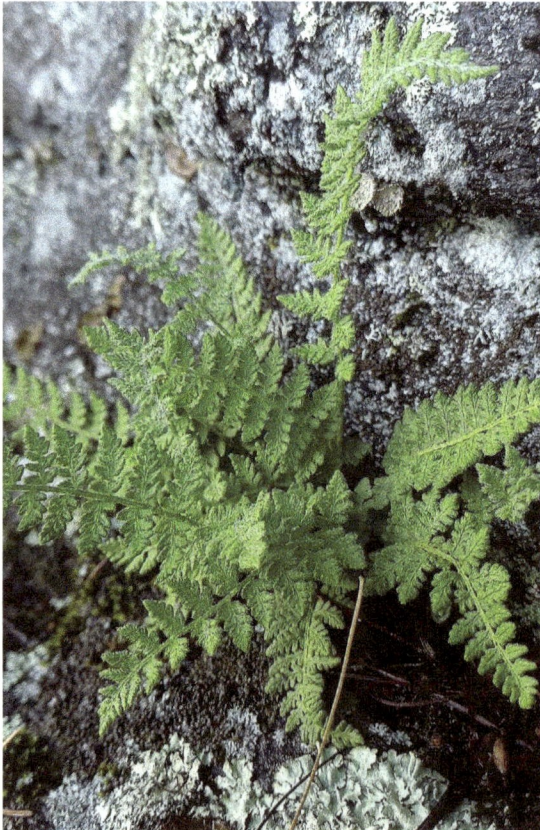

WOODSIACEAE
Woodsia alpina

WOODSIACEAE *Woodsia ilvensis*

KEY D — small ferns, terrestrial, growing in soil, not on rock outcrops

1 Stipe branched once dichotomously, each branch with 3-7 pinnae in one direction only, the outline of the blade fan-shaped, often wider than long.. PTERIDACEAE, page 207
.... *Adiantum*, page 208

1 Stipe not branched dichotomously, the outline of the blade either longer than wide or triangular and about as wide as long.. 2

2 Fronds pinnatifid or 2-pinnatifid, most of the pinnae not fully divided from one another (the rachis winged by leaf tissue for most or all of its length)..................... 3

2 Fronds 1-pinnate, 1-pinnate-pinnatifid, 2-pinnate, or even more divided (the rachis naked for most of its length, or often winged in the upper portion) 6

3 Sporangia borne on an erect stalk that arises at or above ground level from stipe of sterile leaf blade (joining stipe of sterile leaf above the rhizome) ....
.... OPHIOGLOSSACEAE, page 164
.... *Botrychium*, page 165
.... *Botrypus*, page 179

3 Sporangia either borne on normal leaf blades or on specialized (fertile) fronds 4

THELYPTERIDACEAE
Amauropelta noveboracensis

4 Fronds all alike, sori on normal leaf blades ....
.... THELYPTERIDACEAE, page 225

4 Fronds of two types; sori on fronds significantly different than normal fronds 5

5 Fertile frond woody, brown, with bead-like segments; pinnae margins entire, often wavy or the lowermost somewhat lobed; pinnae nearly opposite . ONOCLEACEAE, page 159
.... *Onoclea sensibilis*, page 162

5 Fertile leaf stiff but herbaceous, green, the pinnae linear,

PTERIDACEAE
Adiantum

OPHIOGLOSSACEAE
Botrychium

THELYPTERIDACEAE
Coryphopteris

ONOCLEACEAE
Onoclea

not at all bead-like; pinnae margins finely toothed, otherwise slightly wavy or straight; pinnae nearly alternate . . . .
. . . . BLECHNACEAE, page 72
. . . . *Lorinseria areolata*, page 76

6   Fronds broadly triangular in outline, about as broad as long; sporangia borne on an erect stalk that arises at or above ground level from the stipe of the sterile leaf blade (joining the stipe of the sterile leaf above the rhizome) 7

6   Fronds lance-shaped in outline, much longer than broad; sporangia either borne on normal leaf blades, on slightly dimorphic blades, or on an erect stalk that arises at or above ground level from the stipe of the sterile leaf blade (joining the stipe of the sterile leaf above the rhizome). 8

BLECHNACEAE
Lorinseria areolata

7   Sporangia borne on normal leaf blades . . . .
. . . . CYSTOPTERIDACEAE, page 78
. . . . *Gymnocarpium*, page 92

CYSTOPTERIDACEAE
Gymnocarpium

CYSTOPTERIDACEAE
Gymnocarpium dryopteris

7   Sporangia borne on an erect stalk that arises at or above ground level from the stipe of the sterile leaf blade (joining the stipe of the sterile leaf above the rhizome). . . . .
. . . . OPHIOGLOSSACEAE, page 164
. . . . *Sceptridium*, page 185

OPHIOGLOSSACEAE
Sceptridium

OPHIOGLOSSACEAE
Botrychium minganense

8   Blades 1-8 cm long; sporangia borne on an erect stalk that arises at or above ground level from the stipe of the sterile leaf blade (joining the stipe of the sterile leaf above the rhizome). . . . . OPHIOGLOSSACEAE, page 164
. . . . *Botrychium*, page 165

8   Blades 10-30 (-100) cm long; sporangia either on normal

OPHIOGLOSSACEAE
Botrychium

leaf blades or on slightly modified blades 9

9 Fronds evergreen, dark green, somewhat leathery . . . .
. . . . DRYOPTERIDACEAE, page 102
. . . . *Polystichum*, page 126

9 Fronds light to medium green, herbaceous, deciduous to
semi-evergreen 10

DRYOPTERIDACEAE
Polystichum

10 Sori elongate; leaf blades somewhat dimorphic, the fertile
larger and erect, the sterile smaller and prostrate, the larger
leaf blades 2-4 (-6.5) cm wide . . ASPLENIACEAE, page 43
. . . . *Asplenium platyneuron*, page 52

10 Sori round; leaf blades monomorphic; the larger leaf blades
5-15 cm wide . . . . THELYPTERIDACEAE, page 225
. . . . *Phegopteris*, page 230

ASPLENIACEAE
Asplenium platyneuron

THELYPTERIDACEAE
Phegopteris hexagonoptera

THELYPTERIDACEAE
Phegopteris
hexagonoptera

33

KEY E — medium to large ferns, growing on rock, or on thin soil over rock

1   Fronds vine-like, 0.3-10 m long, the branching dichotomous, 1 branch of each dichotomy terminating in a pair of pinnae, the pinnae often widely spaced (more than 10 cm apart) . . . . LYGODIACEAE, page 154
. . . . *Lygodium palmatum*, page 155

LYGODIACEAE
Lygodium palmatum

1   Fronds not vine-like, 0.3-1 m long, the branching not as described above, the pinnae regularly and more-or-less closely spaced (mostly less than 10 cm apart) 2

2   Fronds 1-pinnate-pinnatifid or less divided, the pinnae entire, toothed, lobed or pinnatifid 3

2   Fronds 2-pinnate or more divided, the pinnae divided to their midribs 8

3   Sori elongate, the indusium flap-like, attached along the side; leaf blades (if more than 30 cm long) less than 7 cm wide . . . . ASPLENIACEAE, page 43
. . . . *Asplenium platyneuron*, page 52

ASPLENIACEAE
Asplenium platyneuron

3   Sori circular or globular, the indusium peltate, kidney-shaped, or cuplike; leaf blades (if more than 30 cm long) more than 5 cm wide 4

4   Fronds 1-pinnate, the pinnae toothed and each with a slight to prominent lobe near the base on the side towards the leaf tip, dark green, somewhat leathery; indusia peltate . . 5

4   Fronds 1-pinnate-pinnatifid, the pinnae pinnatifid, generally lacking a prominent basal lobe, light green to dark green, herbaceous to slightly leathery; indusium either kidney-shaped or cuplike 6

5   Veins anastamosing, rejoining to form a netlike pattern; pinnae 4-25 pairs per leaf; introduced fern, reported from MA and NJ . . . . DRYOPTERIDACEAE, page 102
. . . . *Cyrtomium falcatum*, page 102

DRYOPTERIDACEAE
Cyrtomium falcatum

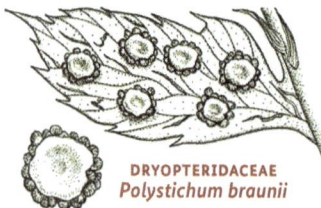

DRYOPTERIDACEAE
Polystichum braunii

5   Veins branching dichotomously, free, not rejoining to form a netlike pattern; pinnae 25-50 pairs on larger fronds; native ferns . . . . DRYOPTERIDACEAE, page 102
. . . . *Polystichum*, page 126

DRYOPTERIDACEAE
Polystichum

6 Vascular bundles in the stipe 3-7 . . . .
. . . . DRYOPTERIDACEAE, page 102
. . . . *Dryopteris*, page 102

DRYOPTERIDACEAE
Dryopteris

6 Vascular bundles in the stipe 2, uniting above 7

7 Indusium kidney-shaped, arching over the sorus . . . .
. . . . THELYPTERIDACEAE, page 225

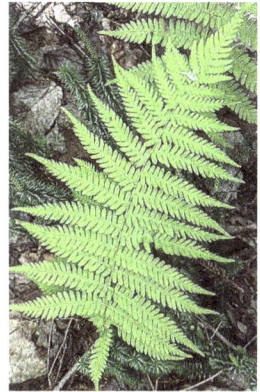

THELYPTERIDACEAE
Amauropelta

7 Indusium cuplike, attached beneath sorus and consisting of 3-6 lance-shaped to ovate segments. . . . .
. . . . WOODSIACEAE, page 236
. . . . *Woodsia*, page 236

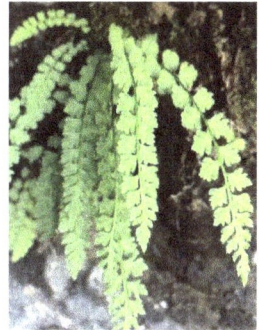

WOODSIACEAE
Woodsia

8 Sori marginal and borne on underside of the false indusium; stipes and rachis shiny black or reddish-black, glabrous except at the very base of the stipe; pinnules fan-shaped or obliquely elongate . . . . PTERIDACEAE, page 207
. . . . *Adiantum*, page 208

8 Sori not marginal, borne on undersurface of leaf blade (or if marginal, as in *Pteridium* and *Dennstaedtia*, borne on undersurface of the leaf); stipes darkened only near base (if at all), rachis green, tan, or reddish; pinnules not notably fan-shaped or obliquely elongate 9

PTERIDACEAE
Adiantum

9 Blades broadly triangular in outline, about as long as wide
.... DENNSTAEDIACEAE, page 96
.... *Pteridium aquilinum*, page 98

9 Blades elongate, mostly lanceolate, generally 4x or more as
long as wide **10**

THELYPTERIDACEAE
Coryphoteris simulata

DENNSTAEDTIACEAE
Pteridium aquilinum

10 Outline of leaf blade narrowed to base, the widest point
more than 7 pinna pairs above the base, the lowermost pinnae 1/4 or less as long as the longest pinnae; rhizomes
long-creeping, the fronds scattered, forming clonal patches
.... THELYPTERIDACEAE, page 225
.... *Coryphopteris*, page 228

10 Outline of the leaf blade slightly if at all narrowed to the
base, the widest point less than 5 pinna pairs from the base,
the lowermost pinnae more than 1/2 as long as the longest
pinnae; rhizomes short-creeping, the fronds clustered, not
forming clonal patches (or with rhizomes long-creeping,
fronds scattered, forming clonal patches in *Dennstaedtia
punctilobula*)..................................... **11**

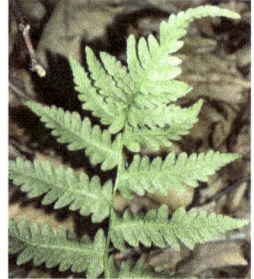

THELYPTERIDACEAE
Coryphoteris

11 Rhizomes long-creeping, fronds scattered, forming clonal
patches; vascular bundles in the stipe 1, U-shaped (even in
the lower stipe); sori very small, marginal in sinuses, the indusium cuplike, 2-parted, the outer part a modified tooth
of the leaf blade; leaf blades conspicuously finely hairy ...
.... DENNSTAEDIACEAE, page 96
.... *Dennstaedtia punctilobula*, page 96

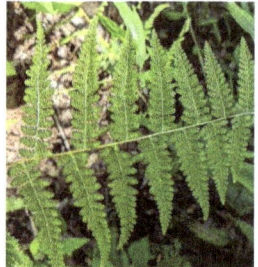

DENNSTAEDTIACEAE
Dennstaedtia punctilobula

11 Rhizomes short-creeping, the fronds clustered, not forming
clonal patches; vascular bundles in the stipe 2-7 (sometimes
uniting to 1 in upper stipe); sori mostly larger, mostly not
marginal, the indusium not as above (though cuplike in
Woodsia obtusa); leaf blades either smooth, with flattened
scales, or finely hairy with gland-tipped hairs.......... **12**

12 Vascular bundles (3-) 5 (-7) in the stipe; mostly larger ferns
of forests............... DRYOPTERIDACEAE, page 102
.... *Dryopteris*, page 102

DRYOPTERIDACEAE
Dryopteris

36

12 Vascular bundles 2 in the stipe (or joining near the leaf blade into 1); ferns of forests and rocky habitats. **13**

13 Fronds 1-pinnate-pinnatifid; indusium cuplike, attached beneath the sorus and consisting of 3-6 lanceolate to ovate segments; mostly smaller ferns on rock . . . .
. . . . **WOODSIACEAE**, page 236
. . . . *Woodsia*, page 236

13 Fronds 2-pinnate-pinnatifid; indusium flaplike or pocketlike, attached at one side of the sorus and arching over it .
. . . . **14**

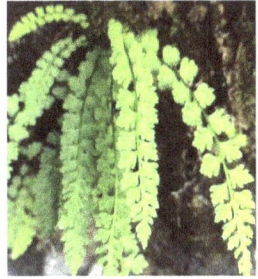

WOODSIACEAE
Woodsia

14 Fronds 10-30 cm wide, the tip acute to acuminate; indusium flaplike . . . . **ATHYRIACEAE**, page 64
. . . . *Athyrium*, page 64

14 Fronds 4-9 cm wide, the tip long-attenuate; indusium pocketlike or hoodlike. . . . . **CYSTOPTERIDACEAE**, page 78
. . . . *Cystopteris bulbifera*, page 80

ATHYRIACEAE
Athyrium angustum

ATHYRIACEAE
Athyrium

CYSTOPTERIDACEAE
Cystopteris bulbifera

KEY F — medium to large ferns, growing in soil (not on rock outcrops)

1 Fronds vine-like, 0.3-10 m long, the branching dichotomous, 1 branch of each dichotomy terminating in a pair of
   pinnae, the pinnae often widely spaced (more than 10 cm
   apart)........................ **LYGODIACEAE**, page 154
   .... *Lygodium palmatum*, page 155

1 Fronds not vine-like, 0.3-3 m long, the branching not as described above, the pinnae regularly and more-or-less closely
   spaced (mostly less than 10 cm apart) **2**

2 Blades broadly (about equilaterally) triangular, pentagonal,
   or flabellate in outline, 0.7-1.3x longer than wide **3**

2 Fronds elongate in outline, mostly ovate, lanceolate,
   oblanceolate, or narrowly triangular, 1.5-10x or more longer
   than wide.. **5**

3 Blades fan-shaped in outline, the stipe branched once di-
   chotomously, each branch bearing 3-7 pinnae ....
   .... **PTERIDACEAE**, page 207
   .... *Adiantum*, page 208

3 Blades broadly triangular in outline, the stipe not branched
   dichotomously **4**

4 Sporangia in a stalked, specialized, fertile portion of the
   blade; texture of mature blades somewhat fleshy; plants
   solitary from a short underground rhizome with thick, mycorrhizal roots; plants of moist forests ....
   .... **OPHIOGLOSSACEAE**, page 164
   .... *Botrypus virginianus*, page 179

4 Sporangia in marginal, linear sori, indusium absent, protected by the revolute leaf margin and a minute false indusium; texture of mature leaf blades hard and stiff; plants
   colonial from deep rhizomes; plants of moist to dry woodlands and openings .... **DENNSTAEDIACEAE**, page 96
   .... *Pteridium aquilinum*, page 98

5 Fronds 2-pinnate or more divided, the pinnae divided to
   their midribs **6**

LYGODIACEAE
Lygodium palmatum

PTERIDACEAE
Adiantum

OPHIOGLOSSACEAE
Botrypus virginianum

OPHIOGLOSSACEAE
Botrypus virginianum

DENNSTAEDTIACEAE
Pteridium aquilinum

5   Fronds 1-pinnate-pinnatifid or less divided; the pinnae entire, toothed, lobed or pinnatifid **10**

6   Blade divided into sterile and fertile portions; sterile pinnae located below terminal fertile pinnae, the sterile pinnules 30-70 mm long and 8-23 mm wide, finely toothed, tip rounded to somewhat pointed; fertile pinnae greatly reduced in size, the fertile pinnules 7-11 mm long and 2-3 mm wide . . . . **OSMUNDACEAE**, page 195
. . . . *Osmunda spectabilis*, page 196

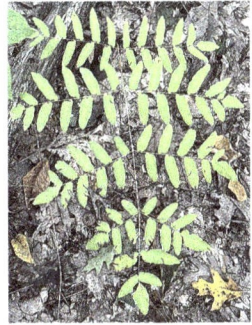

OSMUNDACEAE
Osmunda spectabilis

6   Blade not divided into sterile and fertile portions, the pinnules bearing sporangia only slightly if at all reduced in size, both fertile and sterile pinnules mostly 4-20 mm long and 2-10 mm wide **7**

7   Rhizomes long-creeping, fronds scattered, forming patches; vascular bundles in the stipe 1, U-shaped (even in the lower stipe); sori very small, marginal in sinuses, the indusium cuplike, 2-parted, the outer part a modified tooth of the leaf blade; leaf blades conspicuously finely hairy . . . .
. . . . **DENNSTAEDTIACEAE**, page 96
. . . . *Dennstaedtia punctilobula*, page 96

DENNSTAEDTIACEAE
Dennstaedtia punctilobula

7   Rhizomes short-creeping, the fronds clustered, not forming patches; vascular bundles in stipe 2-7 (sometimes uniting to 1 in upper stipe); sori mostly larger, mostly not marginal, the indusium not as above (though cuplike in *Woodsia obtusa*); leaf blades either smooth, with flattened scales, or finely glandular hairy **8**

8   Vascular bundles (3-) 5 (-7) in the stipe . . . .
. . . . **DRYOPTERIDACEAE**, page 102
. . . . *Dryopteris*, page 102

8   Vascular bundles 2 in the stipe (or joining upwards near blade into 1) **9**

DRYOPTERIDACEAE
Dryopteris

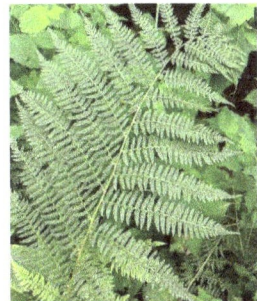

DRYOPTERIDACEAE
Dryopteris cristata

9   Fronds more than 10 cm wide, the tip acute to acuminate; indusium flaplike; pealike bulblets absent . . . .
. . . . **ATHYRIACEAE**, page 64
. . . . *Athyrium*, page 64

ATHYRIACEAE
Athyrium

39

9   Fronds 4-9 cm wide, the tip long-tapering; indusium pock-
    etlike or hoodlike; bulblets often present on upper portion
    of blade . . . . CYSTOPTERIDACEAE, page 78
    . . . . *Cystopteris bulbifera*, page 80

10 Fronds 1-pinnatifid, most of the pinnae not fully divided
    from one another (rachis winged by leaf tissue for most or
    all of its length); fronds dimorphic, the fertile much modi-
    fied, stiff and/or woody **11**

10 Fronds 1-pinnate or 1-pinnate-pinnatifid, the pinnae fully
    divided from one another (rachis naked for most of its
    length, but often winged in upper portion); fronds dimor-
    phic or not. **12**

CYSTOPTERIDACEAE
Cystopteris bulbifera

11 Fertile leaf woody, brown, with bead-like segments; pinnae
    margins entire, often wavy or the lowermost even some-
    what lobed; pinnae mostly opposite along rachis . . . .
    . . . . ONOCLEACEAE, page 159
    . . . . *Onoclea sensibilis*, page 162

ONOCLEACEAE
Onoclea sensibilis

ONOCLEACEAE
Onoclea sensibilis

11 Fertile frond stiff but herbaceous, green, the pinnae linear,
    not at all bead-like; pinnae margins finely serrulate, other-
    wise slightly wavy or straight; pinnae mostly alternate along
    rachis . . . . BLECHNACEAE, page 72
    . . . . *Lorinseria areolata*, page 76

12 Rhizomes long-creeping, fronds scattered, forming patches
    . . . . **13**

BLECHNACEAE
Lorinseria areolata

12 Rhizomes short-creeping, the fronds clustered, not forming
    patches (or rhizomes of both long and short, but fronds
    borne only in clusters on the short erect rhizomes in *Mat-
    teuccia*). **14**

13 Sori roundish, away from the main veins; pinna lobes of
    sterile fronds with the lateral veins free and pinnately
    arranged (the lowermost lateral vein sometimes joining that
    of the adjacent pinna lobe just below the sinus, but the re-
    mainder of the lateral veins all free . . . .
    . . . . THELYPTERIDACEAE, page 225
    . . . . *Thelypteris*, page 234

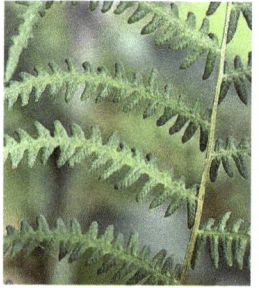

THELYPTERIDACEAE
Thelypteris

13 Sori elongate, end to end along either side of the main veins; pinna lobes of sterile fronds with netted, chain-like venation along the central vein .... BLECHNACEAE page 72
.... *Anchistea virginica*, page 74

14 Plants medium to large, fronds typically 60-300 cm tall; fronds either strongly dimorphic, the fertile fronds very un-like the sterile, brown at maturity (*Matteuccia* and *Os-mundastrum cinnamomea*) or fertile pinnae very unlike the sterile, brown at maturity, borne as an interruption in the blade, with normal green pinnae above and below (*Os-mundastrum claytoniana*); rachis scaleless, stipe scaleless (except at the base in *Matteuccia*) 15

BLECHNACEAE
Anchistea virginica

14 Plants mostly smaller, the fronds 30-100 cm tall (except *Dryopteris celsa* and *D. goldieana* to 15 dm); fronds not at all or only slightly dimorphic, the fertile differing in various ways, such as having narrower pinnae (as in *Polystichum acrostichoides, Diplazium,* and *Thelypteris palustris*) or fertile fronds taller and more deciduous (as in *Asplenium platyneu-ron* and *Dryopteris cristata*), but not as described in the first lead; rachis and stipe variously scaly or scaleless, but at least the stipe and often also the rachis scaly if the plants over 1 m tall 16

15 Fronds strongly tapering to the base from the broadest point (well beyond the midpoint of the blade), lowermost pinnae much less than 1/2 as long as the largest pinnae ..
.... ONOCLEACEAE, page 159
.... *Matteuccia struthiopteris*, page 160

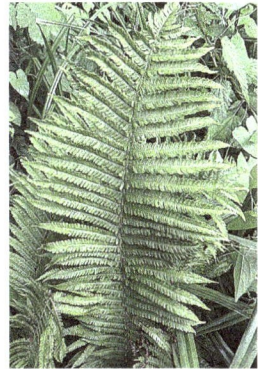
ONOCLEACEAE
Matteuccia struthiopteris

15 Fronds slightly if at all tapering to the base, about equally broad through much of their length, lowermost pinnae much more than 1/2 as long as the largest pinnae ....
.... OSMUNDACEAE, page 195

OSMUNDACEAE

16 Sori elongate, the indusium elongate, attached along one side as a flap 17

16 Sori roundish; the indusium kidney-shaped or nearly round, attached by a central stalk, or sometimes absent........ 19

17 Stipe and rachis lustrous brownish black; fertile fronds 2-8 (-12) cm wide................. ASPLENIACEAE, page 43
.... *Asplenium platyneuron*, page 52

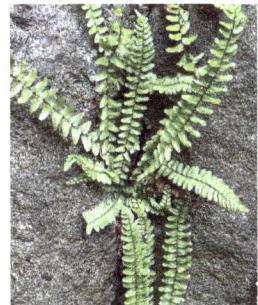
ASPLENIACEAE
Asplenium platyneuron

41

17 Stipe and rachis green; fertile fronds 10-20 (-30) cm wide.
. . . . **18**

18 Fronds 1-pinnate-pinnatifid (the pinnae pinnatifid). . . . . . .
. . . . **ATHYRIACEAE**, page 64
. . . . *Deparia acrostichoides*, page 70

18 Fronds 1-pinnate (the pinnae entire). . . . .
. . . . **DIPLAZIOPSIDACEAE**, page 100
. . . . *Homalosorus pycnocarpos*, page 100

19 Fronds 1-pinnate, the pinnae toothed and each with a slight
to prominent lobe near the base on the side towards the
leaf tip, dark green, somewhat leathery; indusia peltate
(round, stalk attached to the center) . . . .
. . . . **DRYOPTERIDACEAE**, page 102
. . . . *Polystichum acrostichoides*, page 128

19 Fronds 1-pinnate-pinnatifid, the pinnae pinnatifid, generally
without prominent basal lobe, light green to dark green,
herbaceous to somewhat leathery; indusium kidney-shaped
. . . . **20**

20 Vascular bundles in the stipe 4-7 . . . .
. . . . **DRYOPTERIDACEAE**, page 102
. . . . *Dryopteris*, page 102

20 Vascular bundles in the stipe 2, uniting upwards . . . .
. . . . **THELYPTERIDACEAE** page 225
. . . . *Thelypteris*, page 234

ATHYRIACEAE
Deparia acrostichoides

DIPLAZIOPSIDACEAE
Homalosorus pycnocarpos

DRYOPTERIDACEAE
Polystichum acrostichoides

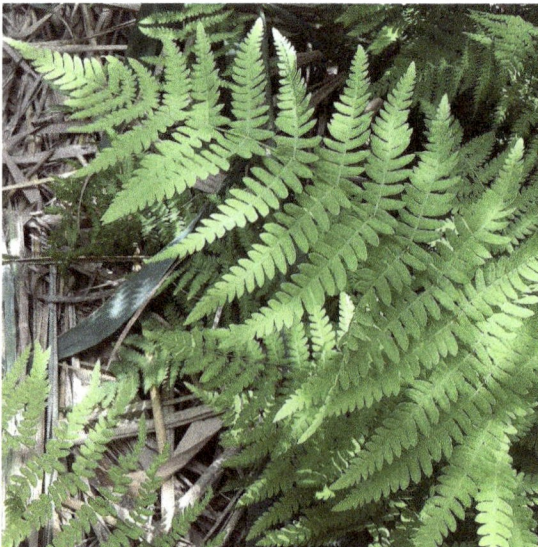

THELYPTERIDACEAE
Thelypteris

ASPLENIACEAE *Spleenwort family*
Asplenium L. SPLEENWORT

SMALL, DELICATE FERNS with short-creeping to erect rootstocks and slender stipes. Worldwide, the family includes two genera and an estimated 730 species (10 species in our flora), most of which occur in tropical and subtropical regions; many species of *Asplenium* are of hybrid origin. In the tropics, spleenworts can be found on the forest floor and also growing in trees (epiphytes). In temperate regions, *Asplenium* are more commonly found on rock in crevices and cracks; a number of our species are restricted to limestone.

KEY CHARACTERS
• Fronds 1-pinnate, all alike or the fertile slightly smaller.
• Linear sori arranged in herringbone fashion along veins.
• Single indusial flap of tissue opening in direction of pinna midrib (except in *Asplenium scolopendrium* with its two-sided, double opening sori).
• Two vascular stipe bundles that unite upwards to form an X-shaped bundle (magnification needed to see this).
• Clathrate scales on rhizome (present on all members of genus, see illus. below). Clathrate, or lattice-like, refers to the scale cells having thick, dark outer walls, surrounding a clear, thin, inner area, similar to the leading between panes of stained glass. However, these can be difficult to see without digging up the plant.

NAME Greek: *a,* not, *splen,* the spleen; refers to early use of a European spleenwort to treat spleen and liver problems.

ADDITIONAL NORTHEASTERN SPECIES AND HYBRIDS
• **Black-stem spleenwort** (***Asplenium resiliens*** Kunze, *right*) is known from southern Pennsylvania (Franklin and Fulton counties), where it occurs on moist to dry outcrops of calcareous rocks. Plants are evergreen, clumped, from a short, upright rootstalk; fronds to 30 cm long; stipe and rachis shiny black. Similar to **ebony spleenwort** (*A. platyneuron*), but in that species, fronds are dimorphic, pinnae margins sharp-toothed, and rachis dark brown, *A. resiliens,* fronds not dimorphic, pinnae nearly entire, and rachis glossy black.

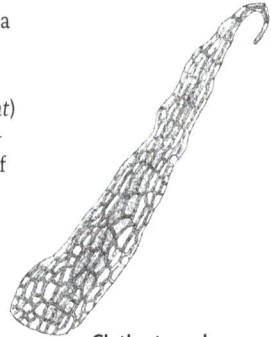

Clathrate scale
Asplenium viride

• *Asplenium* × *clermontiae* Syme; hybrid between *A. rutamuraria* and *A. trichomanes;* reported from Rutland County, Vermont.
• *Asplenium* × *ebenoides* R.R. Scott (pro sp.)] is a sterile hybrid between *A. platyneuron* and *A. rhizophyllum,* found on calcareous rock at scattered locations in the region; fronds are pinnate at base and pinnatifid above.
• *Asplenium* × *gravesii* Maxon (pro sp.); hybrid between *A. bradleyi* and *A. pinnatifidum;* known from Lancaster and York counties in southeastern Pennsylvania.
• *Asplenium* × *trudellii* Wherry (pro sp.); hybrid between *A. montanum* and *A. pinnatifidum;* known from several southern Pennsylvania locations.

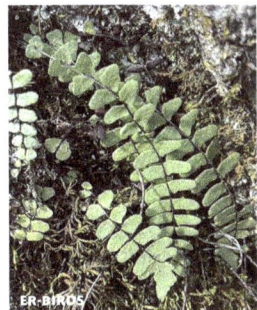

Asplenium resiliens

• **Asplenium × virginicum** Maxon; hybrid between *A. platyneuron* and *A. trichomanes;* reported from Monroe County in eastern Pennsylvania.
• **Asplenium × wherryi** D.M. Sm.; hybrid between *A. bradleyi* and *A. montanum;* reported from Lancaster County and several counties in northern New Jersey.

KEY TO *ASPLENIUM* | SPLEENWORT

1 Leaf blade entire 2
1 Leaf blade pinnately compound. 3
2 Tip of blade long tapering to a gradual point, often with plantlets at the tip, these some-times rooting; regionwide but absent from ME . . . . *Asplenium rhizophyllum*
. . . . WALKING FERN, page 54
2 Tip of blade abruptly pointed to ± rounded, not producing plantlets at the tip; uncommon in NJ, NY . . . . *Asplenium scolopendrium* HART'S-TONGUE FERN, page 58
3 Stipe brown below, green above (same color as rachis) 4
3 Stipe and part or all of rachis brown to black 7
4 Blade long-triangular, lobed, the tip long and slender; NJ, PA . . . *Asplenium pinnatifidum*
. . . . LOBED SPLEENWORT, page 50
4 Blade short-triangular or oblong, 1– to 3-pinnate. 5
5 Blade 1-pinnate; tissue delicate, tardily deciduous; uncommon in ME, NY, VT. . . . .
. . . . *Asplenium viride*, BRIGHT-GREEN SPLEENWORT, page 62
5 Blade 2- or 3-pinnate; tissue firm, evergreen 6
6 Divisions few, pinnae only 3–5 per side, alternate; plants of limestone; CT, MA, NJ, NY, PA, VT . . . . *Asplenium ruta-muraria*, WALL-RUE, page 56
6 Divisions many, opposite; plants of acidic rocks; CT, MA, NJ, NY, PA, RI, VT . . . .
. . . . *Asplenium montanum*, MOUNTAIN SPLEENWORT, page 48
7 Rachis dark only 1/3 its length; pinnae lobed and sharp-toothed, with an auricle on upper side; rare in NJ, PA . . . . *Asplenium bradleyi*, BRADLEY'S SPLEENWORT, page 46
7 Rachis dark throughout 8
8 Sterile and fertile fronds different, the fertile tall and erect, much taller than the spreading sterile fronds; pinnae mostly alternate, the lower much reduced in size; plants usually on soil or sometimes on calcareous rock outcrops; regionwide . . . . . *Asplenium platyneuron*
. . . . EBONY SPLEENWORT, page 52
8 Fronds uniform; pinnae opposite, the lower pinnae little reduced in size; plants usually on rock 9
9 Stipe black; pinnae blue-green, oblong, with a small auricle on upper side of pinnae; mar-gins undulate; rare in southern PA. . . . . *Asplenium resiliens*
. . . . BLACK-STEM SPLEENWORT, page 43
9 Stipe dark brown; pinnae bright green, oval, not auricled; margins round-toothed; region-wide . . . . *Asplenium trichomanes*, MAIDENHAIR SPLEENWORT, page 60

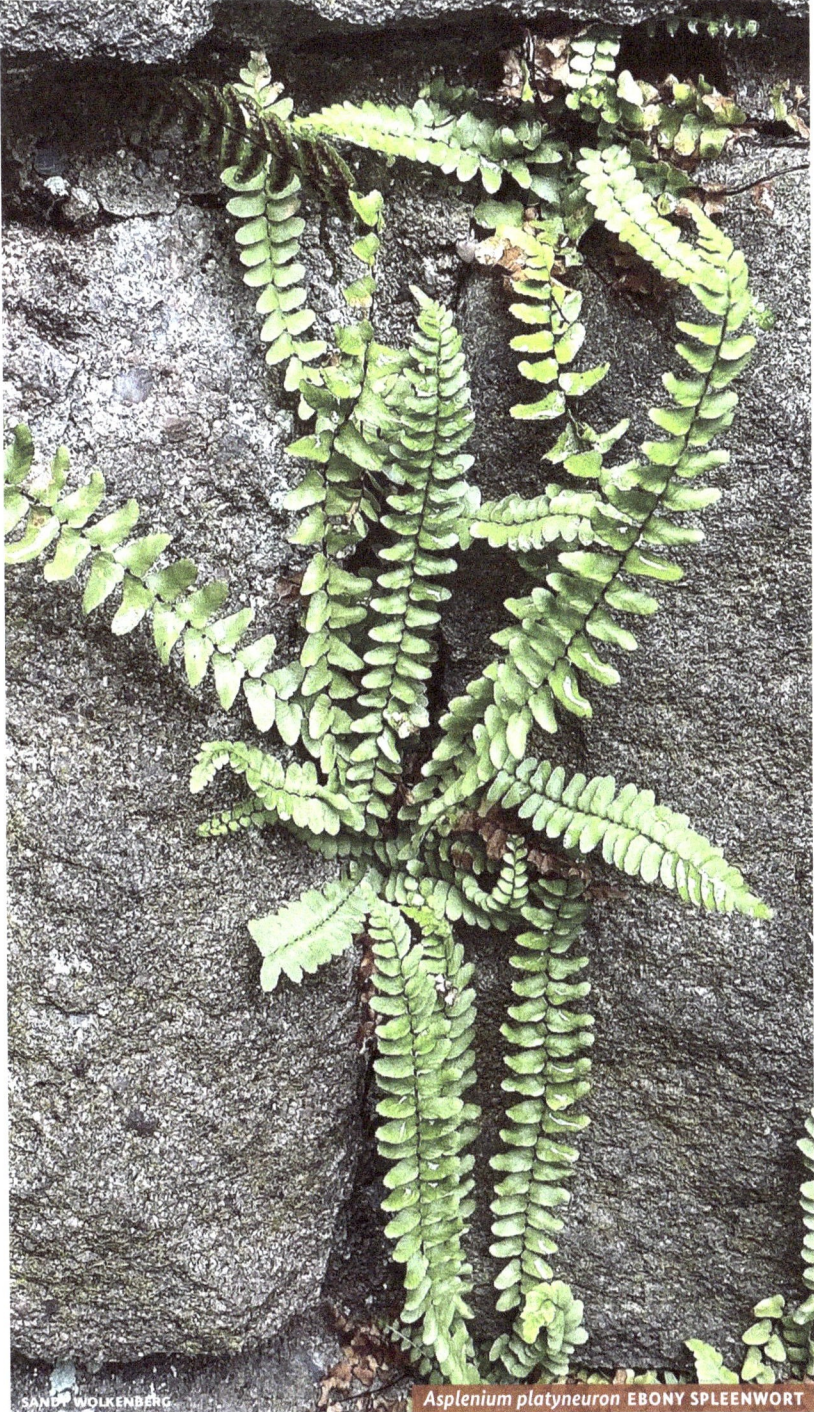

SANDY WOLKENBERG

Asplenium platyneuron **EBONY SPLEENWORT**

Asplenium bradleyi D.C. Eaton
BRADLEY'S SPLEENWORT

DISTRIBUTION NJ | NY *historical* | PA

FIELD TIPS
- small, clumped, evergreen fern
- pinnae dark green, on very short stalks
- stipes dark red-brown, satiny
- crevices of rock cliffs of sandstone (and other non-calcareous rock)
- rare in New Jersey and Pennsylvania

HABITAT

Typically tightly rooted in crevices on bare cliffs of sandstone, granite, chert or other acidic rock; sites usually dry and in full sun. Sometimes found with other spleenworts including *Asplenium montanum, A. pinnatifidum,* and *A. platyneuron,* plus lichens and mosses.

DESCRIPTION

ROOTSTOCK Short-creeping to ascending, occasionally branched.

FROND Evergreen; sterile and fertile fronds alike, 5–30 cm long.

STIPE About half as long as blade, wiry, lustrous dark brown; scales present at base of leaf cluster, these dark reddish to brown, narrowly lance-shaped, grading upwards into hairs.

RACHIS Shiny, dark reddish-brown, the coloration extending to middle of rachis, then fading to green.

BLADE Narrowly lance-shaped in outline, shiny green when young, 1-pinnate-pinnatifid to 2-pinnate, base truncate (not tapered), basal pinnae (or blade divisions) often somewhat smaller.

PINNAE (5-) 8–15 (–20) pairs, alternate or nearly opposite, very short-stalked; upper basal lobe often slightly enlarged and forming an auricle; margins sharp-toothed to jagged-lobed; veins barely evident.

SORI Between margin and midrib, becoming blackish brown, 1–2 mm long; indusium opaque, entire.

SYNONYMS *Asplenium × stotleri* Wherry

CONSERVATION STATUS *endangered* New Jersey, New York (historical records only), Pennsylvania.

SIMILAR SPECIES Often in same habitat as the more common **mountain spleenwort** (*Asplenium montanum*) and **lobed spleenwort** (*A. pinnatifidum*). In *Asplenium montanum,* fronds broader, pinnae on longer stalks, and upper stipe and rachis green and flattened. In *Asplenium pinnatifidum,* pinnae only pinnatifid, in contrast to distinctly pinnate pinnae of *A. bradleyi*.

NATHAN AARON

NOTES Originated as fertile offspring of cross between mountain spleenwort (*Asplenium montanum*) and ebony spleenwort (*Asplenium platyneuron*).

NAME Daniel Eaton named this species in 1873 in honor of Frank H. Bradley, who discovered it in eastern Tennessee.

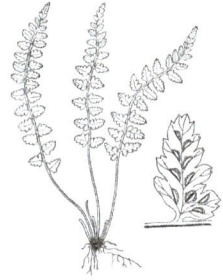

Asplenium bradleyi BRADLEY'S SPLEENWORT

Asplenium montanum Willd.
MOUNTAIN SPLEENWORT

DISTRIBUTION CT | MA | NJ | NY | PA | RI | VT

FIELD TIPS
- small, clumped, evergreen fern
- plants leathery, bluish green
- pinnae widely spaced, stalked; margins indented
- stipe dark purple-brown near base, green above
- rachis green, flattened and winged
- crevices and ledges of rock cliffs, absent from limestone

HABITAT

Moist, shaded crevices and ledges of sandstone, gneiss, granite, and shale; often receiving water from seepage from the surrounding rock; absent from calcareous rocks.

DESCRIPTION

ROOTSTOCK Short-creeping, erect, tightly clumped to produce a cluster of stems; scales dark brown, strongly clathrate (having a lattice-like pattern).

FROND Evergreen; sterile and fertile fronds alike, gradually tapered toward tip and somewhat arched, to 15 cm long and 5 cm wide.

STIPE Shorter than or about same length as blade; brown at base, fading to dull green above; scales at base dark brown, grading into hairs upwards on stipe; with 2 oval bundles at extreme base, united just above base to form one nearly circular bundle.

RACHIS Dull green, winged, flattened; sparsely hairy.

BLADE Oblong-triangular, 2-pinnate at base, always less divided or merely lobed upwards; leathery, dull, often dark blue-green, often with tiny glandular hairs and sparse linear scales.

PINNAE 6 to 9 pairs, opposite to alternate, mostly 3–6 lobed, variable in shape; blunt-tipped; veins forked or simple, never reaching margin, margins coarsely and irregularly indented.

SORI Brown, scattered along veins of pinnules; indusium tan, translucent, jagged on margin, hidden by sporangia at maturity, on one side of the sorus, opening toward the middle of the segment.

SIMILAR SPECIES The dark blue-green color and widely spaced, deeply parted pinnae distinguish *Asplenium montanum* from most other spleenworts.

- The pinnae of **Bradley's spleenwort** (*A. bradleyi*) are toothed and less deeply cut, and the dark color of the stipe continues partway up the rachis in that species.
- **Wall-rue** (*A. ruta-muraria*) has a green stipe, its pinnae have longer petioles and the pinnae are widest near the tip.

• **Brittle bladder fern** (*Cystopteris fragilis*) is more dissected (pinnules present) and oval rather than linear sori.

CONSERVATION STATUS *endangered* Connecticut, Rhode Island; *threatened* New York, Vermont.

NOTE *Asplenium montanum* readily forms hybrids with a number of other *Asplenium* species.

NAME Sent to Willdenow from the North Carolina mountains and named by him in 1810. From Latin, *montanus*, growing on mountains, referring to its preferred habitat.

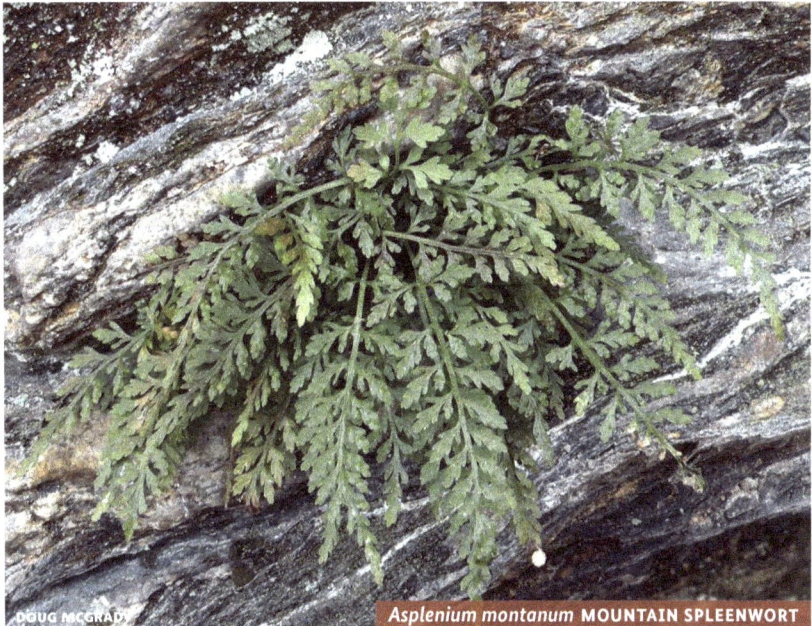

CHOESS

DOUG MCGRADY

Asplenium montanum **MOUNTAIN SPLEENWORT**

Asplenium pinnatifidum Nutt.
LOBED SPLEENWORT

DISTRIBUTION NJ | PA

FIELD TIPS
- small, clumped, evergreen fern
- fronds bright green, crinkled, deeply lobed, the tip
  tail-like and usually pointing downward
- stipe purple-brown at base; rachis green, flat
- crevices in acidic rock cliffs

HABITAT

Dry to moist, shaded crevices of acidic (rarely circumneu-tral) rocks, especially sandstone and gneiss.

DESCRIPTION

ROOTSTOCK Short, very chaffy.

FROND Evergreen, rather tough, sterile and fertile fronds alike, broadly lance-shaped, pinnatifid, deeply cut into pinna-like lobes, long-tapering at tip.

STIPE Dark purple-brown near base, green upwards, 2/3 as long as blade.

RACHIS Flat, green.

BLADE Elongate-triangular in outline, bilaterally somewhat unsymmetrical, tapered to a tail-like tip (and rarely root-ing at tip); pinnatifid (deeply lobed, but not fully pinnate).

PINNAE SEGMENTS vary greatly in size and shape; generally ovate, margins rounded or lobed; veins mostly free, or a few joined.

SORI Numerous, straight or slightly curved, running together with age; indusium membranous, attached at one side.

SYNONYM *Asplenosorus pinnatifidus* (Nutt.) Mickel

CONSERVATION STATUS *endangered* New Jersey.

SIMILAR SPECIES Somewhat similar to its parent *Asplenium rhizophyllum*, but *A. pin-natifidum* distinctly lobed when mature, tends to have longer stipes in proportion to its leaf size, and has a more upright habit; also, lobed spleenwort grows in acid soil and is typically a solitary plant; *Asplenium rhizophyllum* prefers limestone and forms colonies.

NOTE Lobed spleenwort originated as a hybrid between *Asplenium montanum* and *A. rhizophyllum.*

NAME Discovered by Muhlenberg in Lancaster County, Pennsylvania.

CHOESS

Asplenium pinnatifidum LOBED SPLEENWORT

Asplenium

Asplenium platyneuron (L.) B.S.P.
EBONY SPLEENWORT

DISTRIBUTION CT | MA | ME | NH | NJ | NY | PA | RI | VT

FIELD TIPS
- densely clumped fern growing on both soil and rock
- fronds of two types, the sterile fronds forming rosette at base of larger, darker, upright fertile fronds
- stipe and rachis dark brown, shiny
- pinnae alternate, with ear on upper side
- usually where acidic and at least partially shaded

HABITAT

A wide variety of habitats: in partial shade in open woods, old fields and clearings (especially in sandy or loamy soils), less commonly in moss or very shallow soil over rocks, or on limestone or sandstone cliffs and ledges; may also colonize older building foundations and mortared joints.

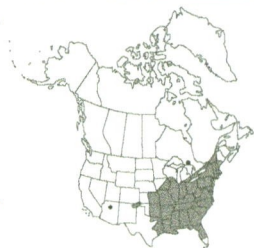

DESCRIPTION

ROOTSTOCK Erect, short, bearing the fronds in dense tufts; scales sparse, dark brown to black, clathrate (with lattice-like pattern).

STERILE FRONDS Evergreen, prostrate and spreading, to 10 cm long, in a rosette near the ground; more numerous than fertile fronds.

FERTILE FRONDS Semi-evergreen, dark green, longer and more upright than sterile frond; to 40 cm tall and 4 cm wide.

STIPE Much shorter than blade, erect; at first green, then reddish-brown, shiny all the way to end of rachis; threadlike scales present near base, becoming smooth upwards.

RACHIS Smooth, red- or purple-brown, shiny.

BLADE (fertile) linear lance-shaped, widest above the middle, tapering to either end, 1-pinnate, shiny, commonly with tiny glandular hairs and a few linear scales.

PINNAE 25 to 45 pinna pairs, alternate along the rachis; with Christmas stocking-like appearance due to an upward (sometimes also downward) pointing auricle (projection) at base of pinna; margins finely toothed; veins forked, never reaching margin; sterile fronds with fewer pinna pairs; lowest pinnae short, triangular.

SORI Positioned along veins of leaflets; aligned diagonally to costa (pinna midrib); indusium linear, 2 mm long, on one side

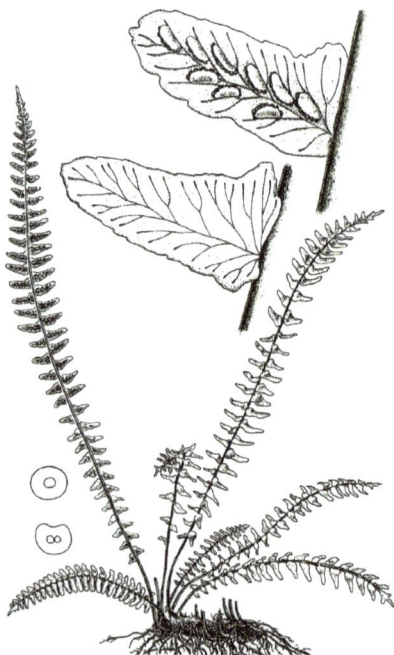

of the sorus, silvery, thin, margins slightly toothed; sporangia brown, maturing in late summer.

SYNONYM *Acrostichum platyneuron* L.

SIMILAR SPECIES **Black-stem spleenwort** (*Asplenium resiliens,* rare in southern Pennsylvania) similar but fronds not dimorphic, pinnae nearly entire, and rachis glossy black. In *A. platyneuron,* fronds dimorphic, pinnae margins somewhat toothed, and rachis dark brown.

NOTE Ebony spleenwort forms hybrids with a number of other spleenworts. The hybrid with mountain spleenwort (*A. montanum*) gives rise to the fertile species Bradley's spleenwort (*Asplenium bradleyi*).

NAME Linnaeus in 1753 proposed the name *Acrostichum platyneuros* for material from Virginia, comprising this plant and a *Polypodium.* Oakes combined the Linnaean species name with the correct genus, as recorded by Eaton in 1878.

NIC TIPPERY

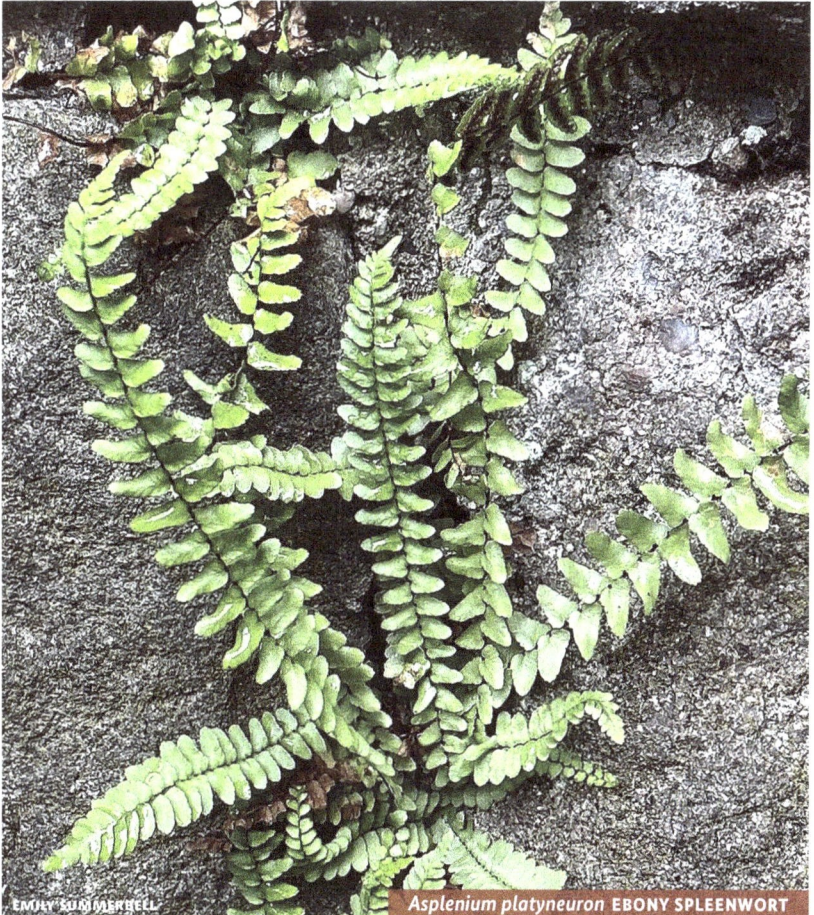
EMILY SUMMERBELL
Asplenium platyneuron **EBONY SPLEENWORT**

Asplenium rhizophyllum L.
WALKING FERN

DISTRIBUTION CT | MA | ME *historical* | NH | NJ | NY | PA | RI | VT

FIELD TIPS
- small evergreen fern
- leaf blades undivided, long-tapering to tip, which can root to form new plants
- found on moist, mossy limestone

HABITAT

Moist, shaded, mossy limestone or dolomite rocks, boulders, ledges and crevices, generally found on northerly exposures; also in moist deciduous woods.

DESCRIPTION

ROOTSTOCK Erect or ascending, scales dark brown, narrowly triangular, clathrate.

FROND Evergreen, sterile and fertile fronds similar but fertile fronds typically larger than sterile fronds; to 30 cm long by 1–5 cm wide.

STIPE Variable in length; dark brown and scaly at base, becoming green and without scales above; upper stipe with tiny club-shaped hairs; vascular bundles 2 at base, these united upwards.

RACHIS Green, dull, nearly smooth, grooved above.

BLADE Simple, leathery, variable in size and shape, narrowly triangular to linear-lance-shaped; basal lobes usually rounded, but sometimes pointed; tip rounded to very long-tapering and, if tapering, often forming plantlet and rooting at tip; sparsely hairy; margins entire to wavy, rarely irregularly lobed; veins obscure, netlike.

PINNAE Absent.

SORI Linear along the veins, scattered irregularly on underside of leaf blade, often joined at vein junction; indusium attached on one side of the sorus, small, green, sporangia gold-yellow when young, becoming red-brown.

SYNONYM *Camptosorus rhizophyllus* (L.) Link

CONSERVATION STATUS *endangered* New Hampshire, Rhode Island.

SIMILAR SPECIES **Lobed spleenwort** (*Asplenium pinnatifidum*) similar but is distinctly lobed, tends to have longer stipes in proportion to its leaf size, and has a more upright habit; also, lobed spleenwort grows in acid soil and is typically a solitary

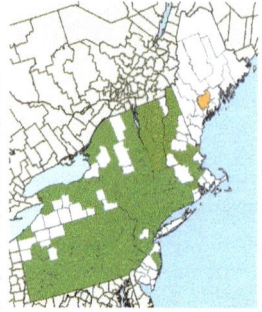

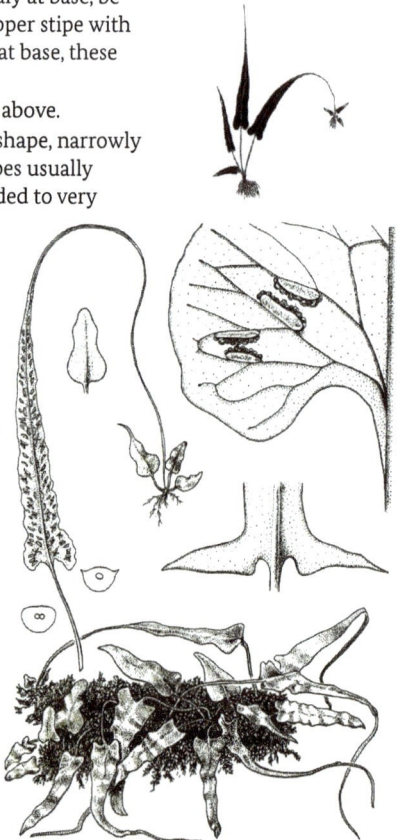

plant; *Asplenium rhizophyllum* prefers limestone and forms colonies.

NAME Linnaeus received this fern from Virginia and Canada and named it *Asplenium rhizophylla* in 1753, the ending *a* being a misprint for *um*. Link founded a genus *Camptosorus* to include this species and a related Asiatic fern in 1833; modern treatments include the species in *Asplenium*.

ELIAS

TOM SCAVO

Asplenium rhizophyllum **WALKING FERN**

Asplenium ruta-muraria L.
WALL-RUE

DISTRIBUTION CT | MA | NJ | NY | PA | VT

FIELD TIPS
- small evergreen fern
- pinnae widely spaced, stalked, divided into small rounded or fan-shaped segments
- shaded limestone cliffs and rocky slopes

HABITAT

Crevices in limestone cliffs and mossy, calcareous talus slopes; in partial to full shade, rarely in full sun; may colonize masonry walls. Best growth made in partial shade on neutral to alkaline soil.

DESCRIPTION

ROOTSTOCK Erect, sparsely branched, tipped with crowded remains of old stipes; scaly, the scales hidden in the roots, dark brown, clathrate (with lattice-like pattern).

Frond Evergreen, clumped, fertile and sterile fronds alike, bluish green, arching and delicate, to 15 cm long by 5 cm wide, somewhat resembling flat-leaved parsley or common rue (*Ruta graveolens*), leading to its scientific name.

STIPE Generally longer than blade, 1.5–7 cm long, green throughout or the base purple-brown; with a few scales at base, upwards becoming multicellular hairs; with two circular or oval bundles at base, united above to form one bundle.

RACHIS Green; smooth, grooved, with sparse hairs

BLADE Oblong-triangular in outline, 2-pinnate at base, always less divided upwards, leathery, dull, usually with tiny glandular hairs and a few linear scales.

PINNAE 2 to 5 pairs, these widely spaced and distinctly stalked, alternate, variable in shape, rounded to fan-shaped; margins facing rachis and costa entire; opposite margin with blunt teeth; serrate or crenate; veins forked, not extending to margin.

SORI Linear, along veins, often covering entire underside; indusium translucent, pale tan, margin fringed with hairs, hidden by sporangia at maturity, on one side of the sorus, opening toward middle of the segment; sporangia brown.

SYNONYMS *Asplenium cryptolepis* Fernald, *Amesium ruta-muraria* (L.) Newman

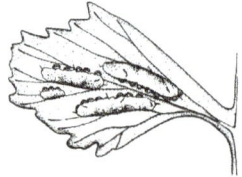

CONSERVATION STATUS *threatened* Connecticut, Massachusetts.

NOTE Our smallest spleenwort and only spleenwort of the region with a triangular-shaped blade (in outline); easily overlooked due to small size and sometimes inaccessible rock cliff habitat; the clumped fronds also have a habit of curling up in dry weather. A common fern on stone walls in Europe.

NAME The European wall-rue was named *Asplenium ruta-muraria* by Linnaeus in 1753.

JACEK PIETRUSZEWSKI

DAVE RICHARDSON

Asplenium ruta-muraria WALL-RUE

Asplenium scolopendrium L.
HART'S-TONGUE FERN

DISTRIBUTION NJ | NY

FIELD TIPS
- distinctive, very rare, evergreen fern
- leaf blades entire, strap-shaped, with smooth to wavy margins
- sori elongated, of varying length

HABITAT

Cool, moist, shaded dolomite boulders and crevices, talus slopes, and rocky woodlands underlain by dolomite.

DESCRIPTION

ROOTSTOCK Erect, short, some tipped with older, dead stipe bases.

FROND Evergreen, fertile and sterile fronds alike, arching, glossy bright green, somewhat leathery, to 35 cm long by 5 cm wide.

STIPE Shorter than blade, 4–12 cm long, 1–2 mm wide, light brown, grooved, with scales when young; vascular bundles 2, C-shaped, joined upwards.

RACHIS Light brown at base, yellowish upwards, smooth.

BLADE Simple, strap-shaped or lance-shaped, uniform in width, base lobed or heart-shaped, heart shaped to lobed at base of blade; tip with blunt point; margins entire becoming wavy with age; veins forked, stopping short of margin.

PINNAE Absent.

SORI Linear, paired along a vein so that diagonal to rachis, in upper half of blade, indusium thin, flap-like, opening toward vein, whitish, on one side of the sorus; sporangia brown.

SYNONYMS *Phyllitis japonica* subsp. *americana* (Fernald) A. & D. Löve, *Phyllitis scolopendrium* var. *americana* Fernald

CONSERVATION STATUS *threatened* New York.

NOTE Common in Europe; rare in North America where populations widely scattered, and given varietal status (*A. scolopendrium* var. *americanum*) because of the smaller fronds, narrower scales, and blades tending to bear sori in the upper half rather than on the entire length. Popular as an ornamental plant, and cultivars available with varying frond form, including frilled margins, forked fronds and cristate forms. The American variety is

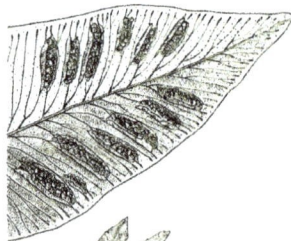

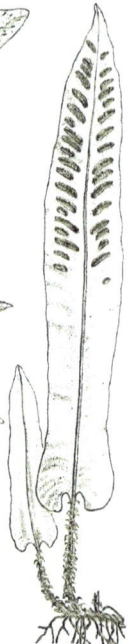

considered to be difficult to cultivate, and cultivated plants are derived from European forms.

NAME The European hart's-tongue was known to the ancients as *Scolopendrium* (from the Greek *scolopendra*, centipede), alluding to the two rows of sori which resemble feet of a centipede; Linnaeus made this the species name under *Asplenium* in 1753. The tongue-shaped leaves lead to the common name; hart is an old word for deer.

AGNIESZKA KWIECIEŃ

BRAD VON BLON

Asplenium scolopendrium HART'S-TONGUE FERN

Asplenium trichomanes L.
MAIDENHAIR SPLEENWORT

DISTRIBUTION CT | MA | ME | NH | NJ | NY | PA | RI | VT

FIELD TIPS
- small, densely clumped, evergreen fern
- sterile fronds small and spreading, fertile fronds longer and upright
- stipe and rachis shiny purple-brown
- pinnae opposite, rounded, not eared
- rock cliffs and boulders, usually where shaded

HABITAT

Damp, shaded, mossy limestone cliffs and boulders, limestone crevices; also found on non-calcareous rocks such as sandstone.

DESCRIPTION

ROOTSTOCK Short-creeping, often branched, scaly near apex, the scales nearly black or sometimes with brown margin, lance-shaped, clathrate.

FROND Evergreen, 20 cm long by 1.5 cm wide, fertile and sterile fronds alike or nearly so, but the sterile fronds developing earlier and oriented along ground surface.

STIPE Much shorter than blade, wiry, dark purple-brown, shiny, sometimes with a few scales; base with 1–2 oval or round bundles. A diagnostic feature (visible with 10× hand lens) is a narrow wing along entire length of the stipe and rachis.

RACHIS Purple-brown, smooth, grooved; old rachises persistent.

BLADE Linear in outline, 1-pinnate, widest above middle, tapering to either end, dark green, smooth or minutely hairy.

PINNAE 20 to 35 pairs, opposite or nearly so or sometimes alternate, oval, the base unequal and sometimes wedge-shaped, slightly toothed on sides and the rounded apex; veins evident.

SORI Oblong to linear, 2–5 pairs per pinna, situated on the veins between the midrib and the margin; indusium translucent, pale tan, hidden by sporangia at maturity, on one side of the sorus, sporangia brown.

SIMILAR SPECIES

- **Ebony spleenwort** (*Asplenium platyneron*) has shiny, purple-brown stipes and dark green pinnae as in *A. trichomanes*. However maidenhair spleenwort is smaller overall, with once-cut fronds, and very small, oval-shaped pinnae.

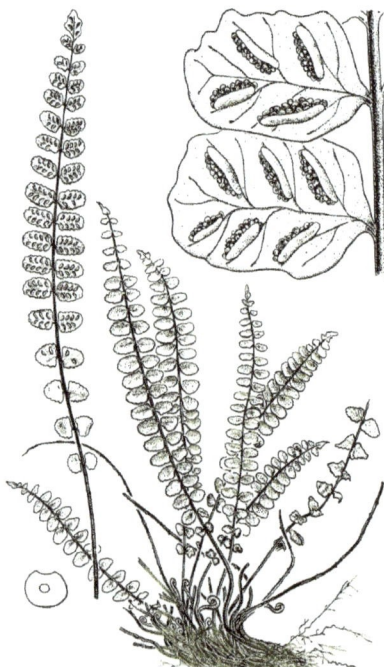

- Distinguished from **rock polypody** (*Polypodium virginianum*) by its prominent, dark rachis, and narrow fronds of nearly constant width.

NOTE Two subspecies, are defined: subsp. *quadrivalens* D. E. Mey., which prefers calcareous substrates, and the more common subsp. *trichomanes,* which prefers acidic substrates.

NAME Named by Linnaeus from European specimens in 1753. Subsequently found to be widespread and essentially identical in North America. From Greek, *thrix,* hair, referring to persistent stipes and rachises without pinnae.

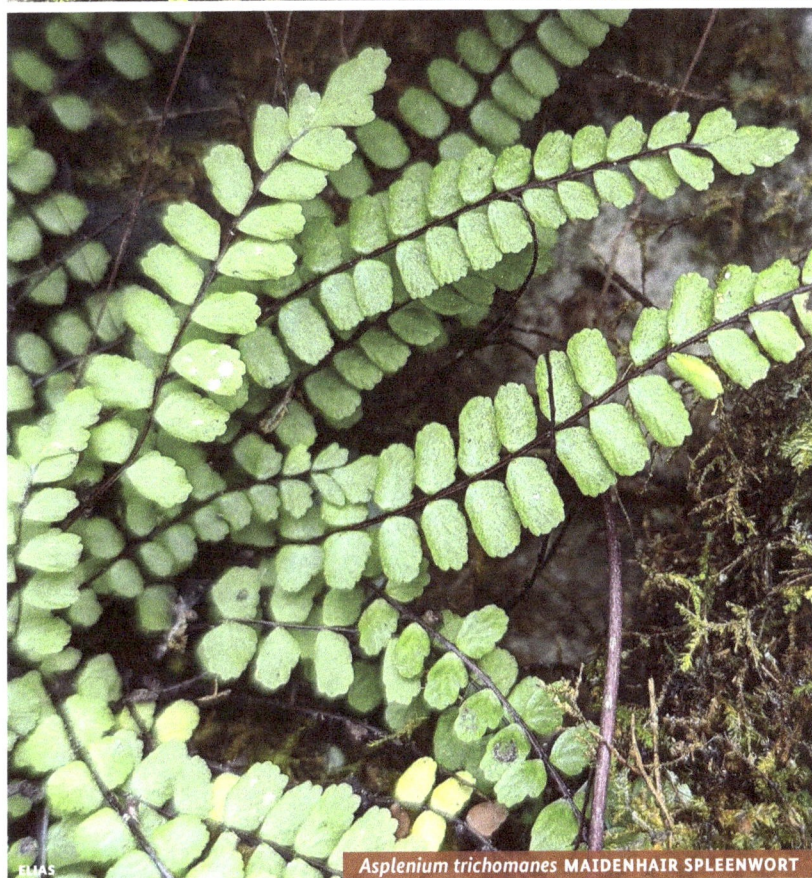

Asplenium trichomanes MAIDENHAIR SPLEENWORT

Asplenium viride Huds.
BRIGHT-GREEN SPLEENWORT

DISTRIBUTION ME | NY | VT

FIELD TIPS
- delicate, clumped, nearly evergreen fern
- lower and middle pinnae widely spaced
- rachis bright green
- shaded limestone habitats

HABITAT

Shaded crevices and ledges of limestone; talus slopes and boulders.

DESCRIPTION

ROOTSTOCK Short-creeping to erect, branching; scales few, dark brown to black, with lattice-like pattern.

FROND Nearly evergreen, clumped from the short rhizome, 5–15 cm long, slightly dimorphic, the fertile fronds stiff, erect; sterile fronds smaller, prostrate.

STIPE Shorter than blade, red-brown at base, green above near rachis, lustrous; usually with a few scales near base, grading into gland-tipped hairs upwards; vascular bundles 1 or 2.

BLADE linear, widest above middle, tapering to either end, 1-pinnate, thin, pale green, smooth or with sparse tiny hairs.

PINNAE 6 to 21 pairs; rhombic or ovate, rounded to acute at tip, subopposite; midrib indistinct; margins round-toothed or entire; veins sometimes forked.

SORI Linear, 1–2 pairs per pinna, paired across the midrib and often joined when mature; indusium white or translucent, entire, often deciduous, attached on one side of the sorus, sporangia brown.

SYNONYMS *Asplenium ramosum* L., *Asplenium trichomanes-ramosum* L.

CONSERVATION STATUS *endangered* Maine, New York; *threatened* Vermont.

SIMILAR SPECIES This fern with its bright green rachis and upper stipe is usually easily identified. Sometimes growing with **maidenhair spleenwort** (*Asplenium trichomanes*) but rachis in that species is purple-brown. Small specimens may resemble **smooth cliff fern** (*Woodsia glabella*), but *Asplenium viride* has elongate sori (vs. round in *Woodsia*), with the indusium attached on one side, typical for the genus.

NAME Described by Linnaeus in his 1753 Species Plantarum, under the name *Asplenium trich. ramosum*. That name was

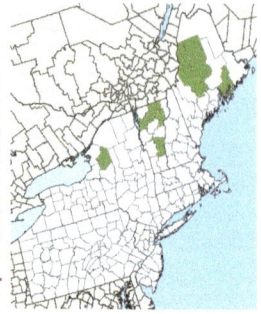

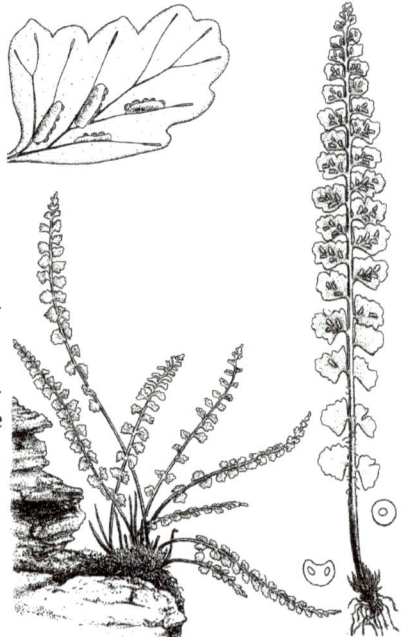

eventually rejected in favour of William Hudson's later name *Asplenium viride; viride* means green.

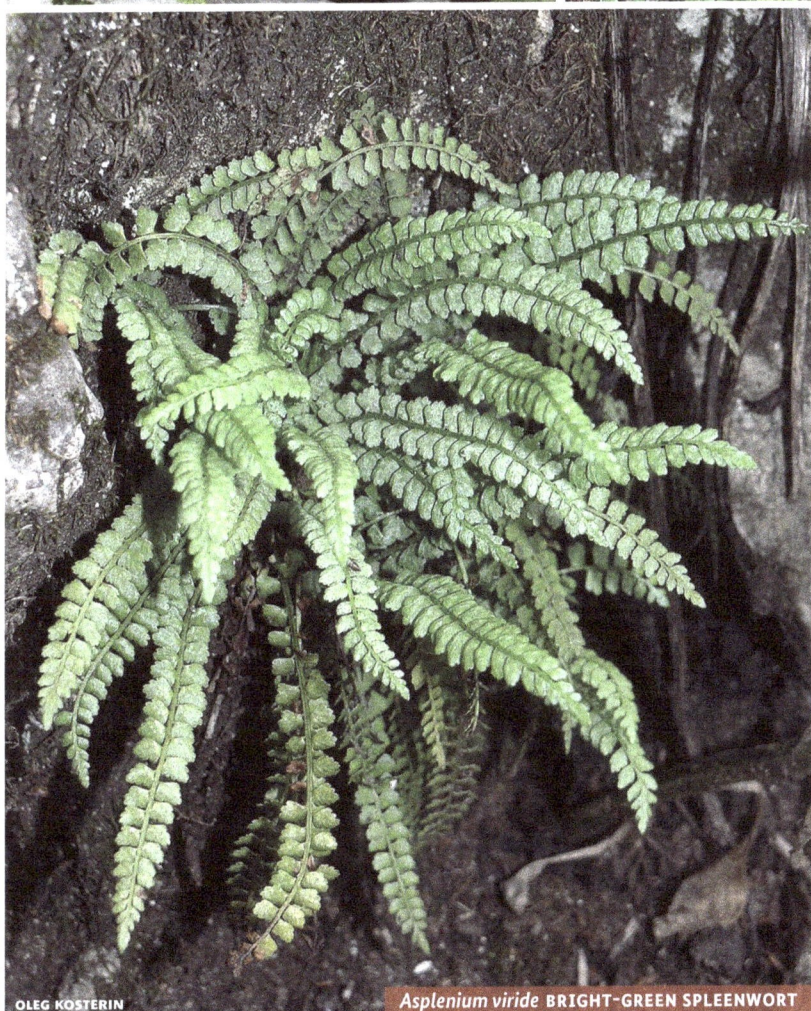

Asplenium viride BRIGHT-GREEN SPLEENWORT

ATHYRIACEAE   *Lady fern family*

MEDIUM TO LARGE FERNS of worldwide distribution; family includes genera and an estimated 650 species (previously mostly included in Aspleniaceae, Dryopteri-daceae or Woodsiaceae); three genera in our flora.

NOTE Until 1800, *Athyrium* was included in genus *Asplenium,* as the indusia are similar. *Deparia* and *Homalosorus,* formerly considered members of Athyrium, are now treated as separate species (and *Homalosorus* now placed in its own family), although the differences are not great. In contrast to *Deparia, Athyrium* has a more deeply grooved rachis, which is continuous from rachis to costa (vs. discontinuous in *Deparia*); in *Homalosorus,* the leaf blade is merely 1-pinnate.

KEY TO ATHYRIACEAE | LADY FERN FAMILY

1   Pinnae margins finely toothed, but otherwise undivided; CT, MA, NH, NJ, NY, PA, VT   . . .
. . . . see *Homalosorus pycnocarpos*, GLADE FERN, page 100
1   Pinnae deeply lobed or divided, margins sometimes finely toothed 2
2   Blades 2-pinnate (with the pinnae again divided), the pinnules also sometimes deeply lobed; common, regionwide. . . . . *Athyrium*, LADY FERN, page 64
2   Blades with deeply lobed pinnae; common, regionwide . . . . *Deparia acrostichoides*
. . . . SILVERY-SPLEENWORT, page 70

Athyrium Roth LADY FERN

MEDIUM TO LARGE FERNS, usually in woodlands. *Athyrium* includes about 230 species worldwide, mostly in temperate regions. There are two native species in our flora.

KEY CHARACTERS

• Clumped deciduous fern.
• Sterile and fertile fronds alike.
• Blade 2-pinnate-pinnatifid (but variable and less and more divided occur).
• Vascular bundles in stipe two only, these uniting upwards to form a U-shape.
• Sori elongate, near base of pinnules, usually hooked at one end or crescent-shaped, covered by persistent indusium.

NAME From Greek, *athyrium,* meaning a small door, the indusium is hinged on one side.

NOTE *Athyrium angustum* one of the region's more common ferns. Circumboreal in distribution, this lady fern is found in North to South America, Europe, and Asia. Lady ferns were historically used to treat a variety of medical conditions. Powdered extracts from the rhizome of *Athyrium filix-femina* (found in Eurasia and western USA), like similar preparations from male fern (*Dryopteris filix-mas*), were used for many ailments, especially for purging worms from the digestive systems of humans and animals.

ADDITIONAL SPECIES The introduced, ornamental **Japanese painted fern [*Athyrium niponicum* (Mett.) Hance, *right*]** is reported from gardens and walkways in Washington

DAVID STANG

County, Rhode Island; it is characterized by variegated leaf blades. It and a large number of cultivated lady fern varieties are used for landscaping; best growth is made in partial shade on moist, acidic soil.

KEY TO *ATHYRIUM* | LADY FERN

1 Petiole scales brown to black-brown; leaf blades elliptic, narrowed to the base, broadest near or just below the middle; leaflets sessile or short-stalked; indusia fringed with non-glandular hairs; common, regionwide *Athrium angustum*, NORTHERN LADY FERN, page 66

1 Petiole scales light brown to brown; leaf blades broad-lanceolate to lanceolate, only slightly narrowed to the base, broadest just above the base; leaflets generally stalked; indusia fringed with glandular or non-glandular hairs; spores dark brown; southern portions of region, absent from ME, NH, VT *Athyrium asplenioides*, SOUTHERN LADY FERN, page 68

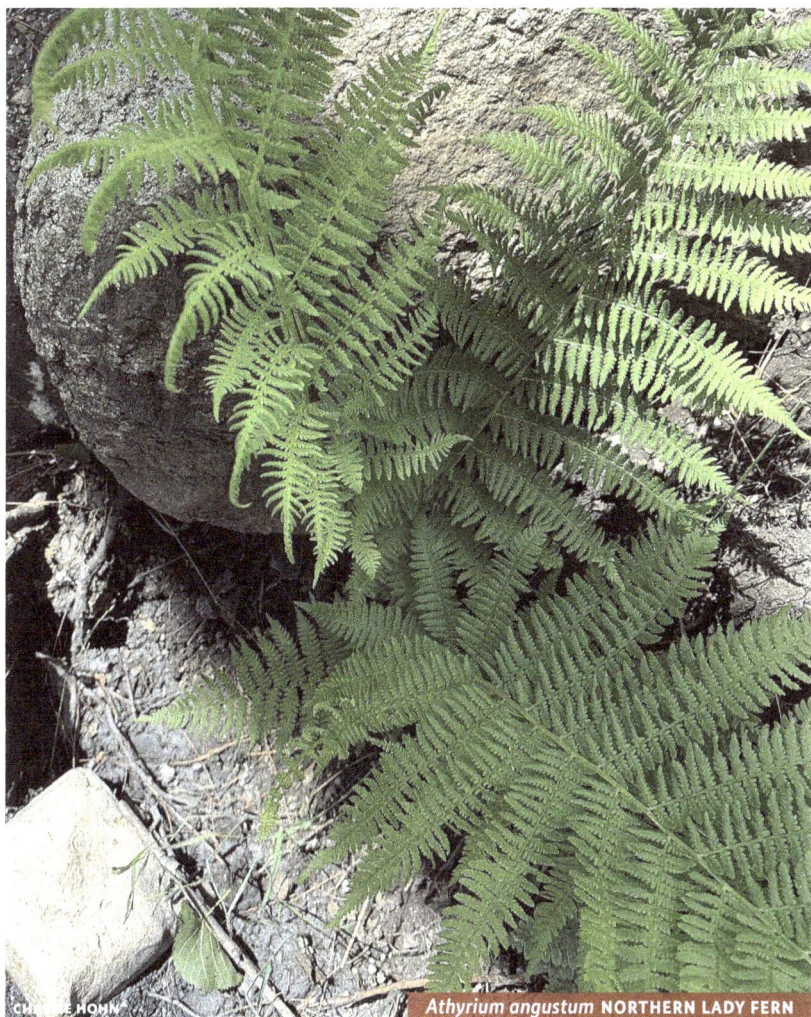

Athyrium angustum NORTHERN LADY FERN

Athyrium angustum (Willd.) K. Presl
NORTHERN LADY FERN

DISTRIBUTION CT | MA | ME | NH | NJ | NY | PA | RI | VT

FIELD TIPS
- common, medium to large fern found across region
- fronds clumped, deciduous
- sori elongate; hooked at one end, horseshoe-shaped, or sometimes straight
- usually in woodlands

HABITAT

Moist woods, wetland margins, moist to wet open areas.

DESCRIPTION

ROOTSTOCK Short-creeping to nearly erect, scaly.

CROZIER Round-oblong, 1–2 cm wide, densely covered with linear, dark brown to purple scales.

FROND Deciduous, forming an irregular clump; new fronds produced all summer; sterile and fertile fronds alike, to 120 cm long by 30 cm wide.

STIPE Straw-colored above, base dark red-brown or black, swollen; in some forms very red, old stipes persistent; scales light to dark brown, lance-shaped; vascular bundles 2, curved, back-to-back, uniting to a U-shape above.

RACHIS Pale, slightly grooved to flat; smooth or with glands or hairs.

BLADE Elliptic to lance-shaped, 2-pinnate-pinnatifid (but variable, less and more divided blades also occur), yellow-green to bright green.

PINNAE 20 to 30 pairs, mostly alternate, stalk very short or absent, oblong lance-shaped, lower pinnae longer or shorter than middle pinnae; costae grooved above, the groove continuous from rachis to costae to costules; margins toothed or lobed; veins free, forking, reaching the margin.

SORI At base of leaf segments, straight, or more often hooked at one end or horse-shoe-shaped; indusium elongate, laterally attached, finely toothed; sporangia brownish gray to dark brown.

SYNONYM *Athyrium filix-femina* subsp. *angustum* (Willd.) Clausen

SIMILAR SPECIES
- Blades of **southern lady fern** (*Athyrium asplenioides*) typically widest near base vs. widest near middle in *Athyrium angustum*.

• **Hay-scented fern** (*Dennstaedtia punctilobula*), not in clumps, fronds with white, sticky gland-tipped hairs.
• **Silvery-spleenwort** (*Deparia acrostichoides*) has blades 1-pinnate-pinnatifid vs. 2-pinnnate-pinnatifid in *Athyrium;* also groove in rachis and costa not continuous in *Deparia*.
• **New York fern** (*Thelypteris novaboracensis*) margins not toothed.
• **Wood ferns** (*Dryopteris* spp.) have round sori, rather than the elongate or curved sori of *Athyrium*.

NAME Linnaeus first named the European lady-fern *Polypodium filix femina*. Michaux altered this to *Nephrodium filix femina* for a related Canadian plant. Other specimens from Canada were named *Aspidium angustum* by Willdenow in 1810; his species was classified as *Athyrium* by Presl in 1825. From the Latin *filix*, fern; *femina*, female or woman.

BEN ARMSTRONG

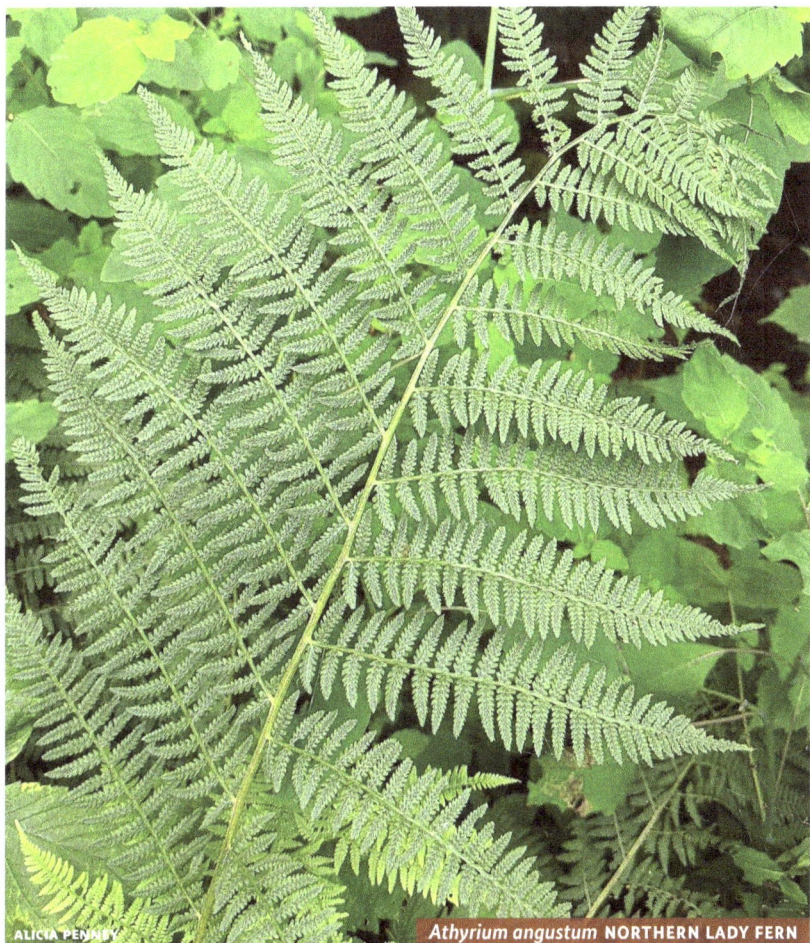
ALICIA PENNEY
Athyrium angustum NORTHERN LADY FERN

Athyrium asplenioides (Michx.) A.A. Eaton
SOUTHERN LADY FERN

DISTRIBUTION CT | MA | NJ | NY | PA | RI

FIELD TIPS
- medium to large fern in southern portion of region
- fronds clumped, deciduous
- sori elongate; hooked at one end, horseshoe-shaped, or sometimes straight
- usually in woodlands

HABITAT

Moist to wet woods, swamps, shores, meadows, thickets, ravines.

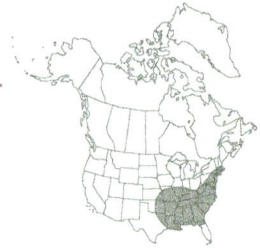

DESCRIPTION

ROOTSTOCK Short-creeping to nearly erect, scaly.

FROND Deciduous, forming a clump; new fronds produced all summer; sterile and fertile fronds alike.

STIPE with light brown or brown, lanceolate scales.

RACHIS glabrous or scaly or with pale glands.

BLADE ovate-lanceolate to lanceolate, 2-pinnate-pinnatifid, usually broadest just above base.

PINNAE usually stalked, oblong-lanceolate to lanceolate, base truncate, apex acuminate.

SORI elongate, straight or hooked at end or horseshoe-shaped; sporangia yellow, their stalks with glandular hairs; indusia ciliate, hairs glandular or nonglandular.

SYNONYMS *Athyrium filix-femina* subsp. *asplenioides* (Michx.) Hultén

CONSERVATION STATUS *threatened* New York.

SIMILAR SPECIES **northern lady fern** (*Athyrium angustum*) similar and is more common in the northeast region, with blades widest near middle; *Athyrium asplenioides* has a more southerly distribution, and its blades are typically widest near base.

DAX LEDESMA

MJPAPAY

EVAN RASKIN

Athyrium asplenioides SOUTHERN LADY FERN

Deparia Hook. & Grev. SILVERY-SPLEENWORT

FERNS OF TERRESTRIAL HABITATS; about 70 species worldwide, primarily in tropical regions of Asia, Africa, Australia and the Pacific Islands; two species in North America; one in our flora.

KEY CHARACTERS

• Medium to large deciduous fern, forming loose clusters from a creeping rhizome.
• Sterile and fertile fronds slightly different, fertile fronds longer, developing earlier than sterile.
• Pinnae widely separated, deeply lobed but not cut into pinnules.
• Sori in herringbone pattern.
• Indusia linear, whitish when young.

SIMILAR GENERA Distinguished from **lady fern** (*Athyrium*) and **glade fern** (*Homalosorus*) by discontinuity in the grooves from rachis to costae, and by the hairs on the rachis and costae. The linear sori are very similar to *Homalosorus*, but are never back-to-back along a vein.

NAME Previously included in *Asplenium, Athyrium* and *Diplazium.* From the Greek *depas,* cup or a beaker, thought to describe shape of the indusium.

Deparia acrostichoides (Sw.) M. Kato
SILVERY-SPLEENWORT

DISTRIBUTION CT | MA | ME | NH | NJ | NY | PA | RI | VT

FIELD TIPS

• medium to large deciduous fern, sterile and fertile fronds slightly different
• fronds single but clustered from rhizome
• blade tapered at both ends; pinnae widely separated and deeply lobed; lowest pair of pinnae point downwards
• sori in herringbone pattern
• indusia linear, whitish when young

HABITAT

Moist, shaded deciduous or mixed conifer-hardwood forests, ravines and slopes, shaded streambanks and seeps, swamp edges; soils typically acidic.

DESCRIPTION

ROOTSTOCK Short-creeping, black, scaly; stipes about 1 cm apart, forming an asymmetric clump.

FROND Deciduous, arising separately (but close together) from the rhizome; sterile and fertile fronds somewhat different;

ASHER HIGGINS

fertile fronds longer, more erect, developing earlier than sterile; to 110 cm long by 20 cm wide.

STIPE 1/2 to 2/3 as long as blade, base dark red-brown, scaly and swollen, straw-colored then green above, the scales changing to long white hairs upwards on stipe; shallowly grooved above; vascular bundles 2, curved or peanut-shaped, uniting to a U-shape above.

RACHIS Light green, very hairy, with a few scales; grooved.

BLADE Elliptic to ovate-lance-shaped, tapering to both ends, 1-pinnate-pinnatifid.

PINNAE 20 to 25 pairs, mostly alternate, oblong, truncate at base, long-tapering; lowest pinnae shorter than middle pinnae and usually pointing downward; costae shallowly grooved above, discontinuous from rachis to costae, hairs on costae and costules; margins entire or finely toothed; veins free, rarely forked, reaching the margin.

SORI Elongate, ± straight, or slightly curved, in a herring-bone pattern, originating on midvein of pinnule; indusium linear, laterally attached, persistent, whitish when young (leading to the 'silvery' name), becoming dark brown, along the vein and opening outward; sporangia brownish early, later blue-gray.

SYNONYMS *Athyrium thelypterioides* (Michx.) Desv., *Diplazium acrostichoides* (Sw.) Butters

SIMILAR SPECIES **Lady fern** (*Athyrium*) has dissected pinnules (not merely lobed) with toothed margins. Sterile fronds of **cinnamon fern** (*Osmundastrum cinnamomeum*) and **interrupted fern** (*O. claytonianum*) are wider, the pinnae closer together, and the stipes scaleless.

NAME First named *Asplenium acrostichoides* by Swartz in 1801, and two years later *A. thelypteroides* by Michaux, with reference to occurrences in Virginia and North Carolina.

BONNIE SEMMLING — *Deparia acrostichoides* SILVERY-SPLEENWORT

BLECHNACEAE *Chain fern family*

OUR MEMBERS OF THIS FAMILY are two species of chain fern (former *Woodwardia*), now separated under the PPG I classification of 2016 into two genera. Members of this family are found in North and Central America, eastern Asia and the Mediterranean region of Europe, and worldwide, there are 24 genera and an estimated 265 species within the family. Our species are medium to large deciduous ferns of shaded wet woods and acidic bogs. Sterile and fertile fronds similar (*Anchistea virginica*) or different (*Lorinseria areolata*), pinnatifid or 2-pinnatifid; veins joining to form small enclosed areas (areolae), then free to the margin or forming additional areolae. Sori linear or oblong, parallel to the midveins, borne along the veinlets, which form the outer side of the first row of areolae. Indusium persistent, opening on the side adjacent to the midrib.

KEY CHARACTERS

- Coarse deciduous ferns of wet acidic places, in shade (or sun where very wet).
- Deciduous, sterile and fertile blades alike (*Anchistea virginica*) or markedly different (*Lorinseria areolata*).
- Colony-forming from long-creeping rootstock.
- Linear sori in chainlike rows along each side of midvein of pinna or pinnule.
- Indusia attached by their outer margin, opening towards midvein.
- Netted veins along the costae and midveins form areoles (enclosed spaces).

NOTE Apart from the distinctive sori, our two species of chain fern have little in common, and the two species, formerly in *Woodwardia,* are now placed in separate genera. *Anchistea virginica* is monomorphic, net-veined only surrounding the sori, has 7–9 bundles at base of the glabrous stipe; in contrast, *Lorinseria areolata* is dimorphic, net-veined throughout, has 2 vascular bundles, and with scaly stipes.

CULTURE *Lorinseria areolata* is readily grown in a shady garden where the soil is wet to moist and acidic. Leaf blades are an attractive pinkish color when young, becoming a glossy bronze-green when mature. *Anchistea virginica* tends to be invasive but can be used to fill a soggy place or grown in a container.

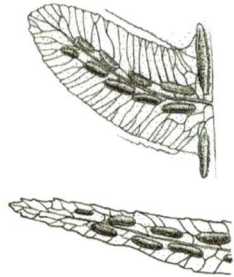

fertile pinna
Anchistea virginica (upper)
Lorinseria areolata (lower)

KEY TO BLECHNACEAE | CHAIN FERN FAMILY

1   Sterile and fertile fronds similar; fronds 1-pinnate-pinnatifid (the pinnae deeply lobed), mostly 13–25 per side; regionwide . . *Anchistea virginica*, VIRGINIA CHAIN FERN, page 74

1   Sterile and fertile fronds very different; sterile fronds merely deeply lobed (the lobes connected at the axis by a narrow wing of tissue), the lobes entire to somewhat undulate, mostly 6–14 per side; fertile fronds with very narrow linear pinnae; southern portions of region . . . . *Lorinseria areolata*, NETTED CHAIN FERN, page 76

JAY HORN

Anchistea virginica VIRGINIA CHAIN FERN

Anchistea virginica (L.) C. Presl
VIRGINIA CHAIN FERN

DISTRIBUTION CT | MA | ME | NH | NJ | NY | PA | RI | VT

FIELD TIPS
- large deciduous fern; sterile and fertile fronds alike
- fronds upright, single from the rhizome, may form colonies with fronds facing one direction
- stipe long, swollen and spongy at base, shiny dark purple-brown
- blade 1-pinnate-pinnatifid
- veins both netted and free
- sori linear, in chainlike rows

HABITAT

Open to shaded acidic swamps, bogs, wet places in woods, sandy or peaty lake shores.

DESCRIPTION

ROOTSTOCK Long-creeping and branching, ropelike; scales few near tips, dark brown, triangular.

CROZIER Reddish brown, scale and hairs absent.

FROND Deciduous, arising singly from the rhizome; sterile and fertile fronds alike, 100 cm long by 30 cm wide.

STIPE About same length as blade; base swollen, spongy, dark purple to black, smooth, with 2 large vascular bundles and 3–7 smaller ones, fewer at top of stipe.

RACHIS Dark purple-brown becoming green upwards; smooth and shiny, grooved.

BLADE Narrowly ovate, widest at middle, 1-pinnate-pinnatifid, thin-textured, with glands and scales when young, the glands persisting.

PINNAE 12 to 23 pairs; alternate, middle ones to 15 cm; lower pinnae bending forward and towards the base; margins finely dentate; veins netted next to the rachis and costae, free otherwise.

SORI In chainlike rows, but distinct along costae; indusium flap-like, opening towards costa; sporangia purple-brown.

SYNONYM *Woodwardia virginica* (L.) Sm.

CONSERVATION STATUS *threatened* Vermont.

SIMILAR SPECIES **Cinnamon fern** (*Osmundastrum cinnamomeum*) but that species grows in clumps rather than in a line from creeping rhizome, has different sterile and fertile fronds, and a green stipe (rather than dark purple-brown).

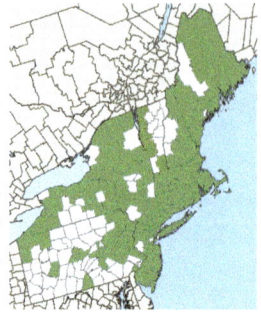

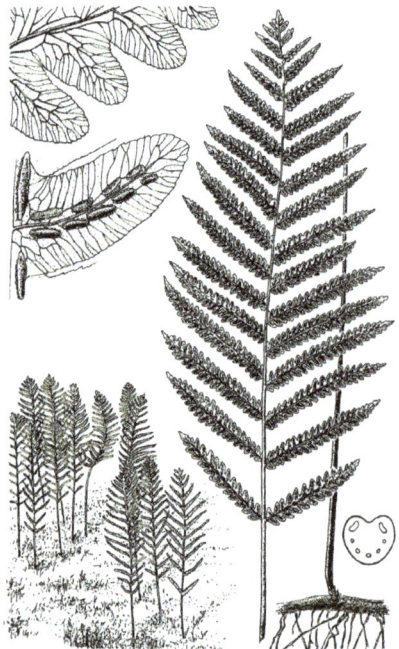

NOTE The chains of sori on the areolae adjacent to the midrib are distinctive.

NAME Collected in Virginia in colonial days, and named *Blechnum virginicum* by Linnaeus. J. E. Smith placed it in genus *Woodwardia* in 1793, but Presl classified it as *Anchistea* in about 1849.

ER-BIRDS

C. J. CHAPMAN   *Anchistea virginica* **VIRGINIA CHAIN FERN**

Lorinseria areolata (L.) C. Presl
NETTED CHAIN FERN

DISTRIBUTION CT | MA | ME *historical* | NH | NJ | NY | PA | RI

FIELD TIPS
- large deciduous fern; sterile and fertile fronds different; fertile pinnae linear
- fronds glossy green, single from rhizome but may form dense colonies
- lowest pinnae pair not winged at rachis
- veins joined to form a network
- sori elongate, in chainlike rows on narrow fertile pinnae

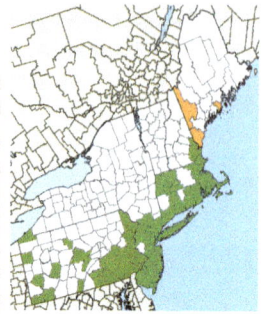

HABITAT

Open to shaded wet woods (where often rooted in mud), bogs, seeps, cobbly beaches, rarely on rock cliffs and ledges; soils typically sandy and acidic.

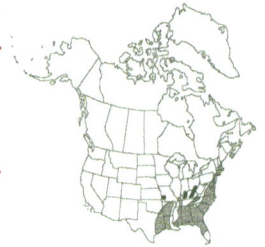

DESCRIPTION

ROOTSTOCK Long-creeping, slender, dark-brown to black; scales many, satiny brown.

CROZIER Densely covered with tan scales.

FROND Deciduous, to 70 cm long, arising singly from the rhizome but may appear massed due to extensive rhizomes; sterile and fertile fronds different, sterile fronds slightly shorter, emerging first, fertile fronds emerging in midsummer, with darker stipes and narrower pinnae. Divisions of fertile frond narrowly linear and almost distinct.

STIPE About same length as blade, with a few scales; sterile fronds reddish brown below, straw-colored above. Fertile fronds darker, to purple-black; vascular bundles 2 at base, merging above to a 3-sided structure.

RACHIS With a few scales on underside, olive-green to straw-colored.

BLADE Lance-shaped, fertile blade 1-pinnate, sterile blade less divided above where it is winged along the rachis; thin-textured, reddish on emergence, bright green later, scaly-glandular when young but soon smooth.

PINNAE 7 to 12 pairs; sterile pinnae alternate, lance-shaped; the veins raised, net-like throughout; fertile pinnae subopposite, thin and contracted; margins wavy and finely toothed.

SORI In chainlike rows on each side of costae; indusium flap-like, tucked under

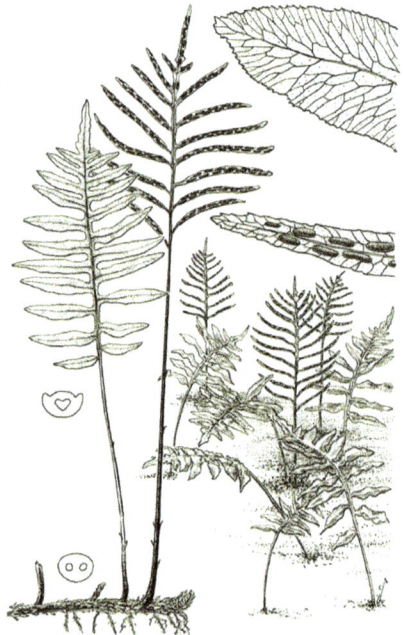

sporangia, disintegrating with age, opening towards the costa; sporangia reddish-brown.

SYNONYMS *Woodwardia areolata* (L.) T. Moore

CONSERVATION STATUS *endangered* New Hampshire, *threatened* Pennsylvania.

SIMILAR SPECIES Superficial resemblance to **sensitive fern** (*Onoclea sensibilis*), but *Lorinseria areolata* has very different fertile fronds (linear sori vs. beads in *Onoclea*), finely toothed rather than entire margins, and by its basal pinnae, which are alternate rather than subopposite as in *Onoclea*.

NOTE Dimorphism sometimes incomplete, resulting in fronds fertile at top and sterile below. *Lorinseria* has been separated from *Woodwardia* in the Pteridophyte Phylogeny Group classification of 2016 (PPG I), on the basis of its anastamosing veins and lobed frond form, as well as its more pronounced frond dimorphism.

NAME This fern reached Linnaeus from Maryland and Virginia, and he named it *Acrostichum areolatum* in 1753. Presl founded the genus *Lorinseria* on it at the same time. Epithet from Latin, *areolatus,* a small place, referring to areas enclosed by the netted veins.

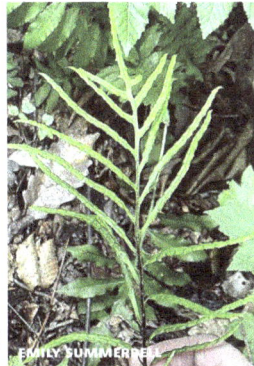

fertile frond EMILY SUMMERBELL

LENA DASHER

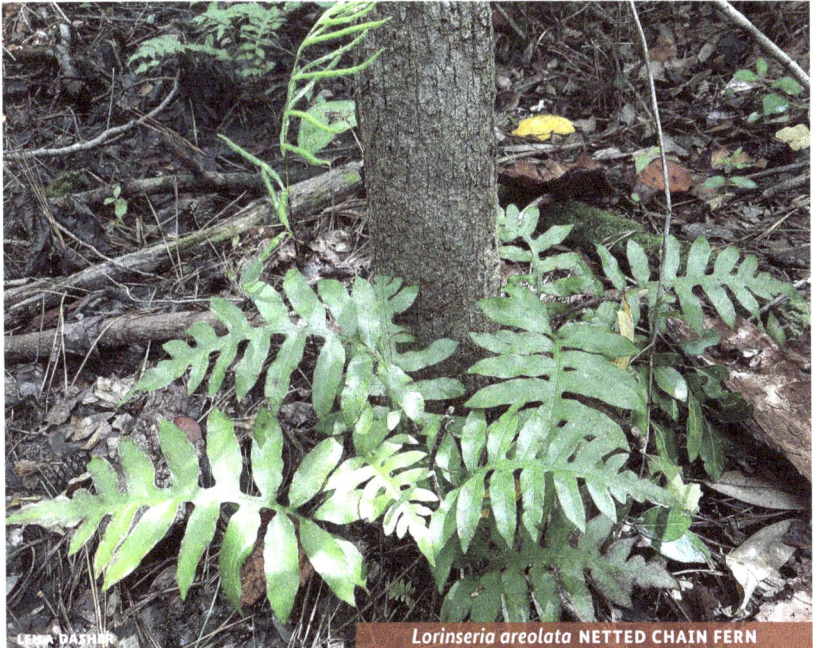

Lorinseria areolata **NETTED CHAIN FERN**

CYSTOPTERIDACEAE *Bladder fern family*

A SMALL FAMILY OF 3 GENERA and an estimated 37 species, mostly in temperate regions. Two genera in our flora, *Cystopteris* and *Gymnocarpium,* common ferns throughout the north temperate zone. *Cystopteris* (at least 26 species, 6 in our flora) also ranges southward in montane habitats of the Andes and Himalayas, and to Australia, New Zealand, Hawaii, and southern Africa. *Gymnocarpium* species number 8 worldwide; 2 in our flora. Characteristic are a swollen knob at the base of each pinna, the indusium is always absent, and leaf-blades 3-parted.

KEY TO CYSTOPTERIDACEAE | BLADDER FERN FAMILY

1 Blade 1-pinnate; indusium present . . . . *Cystopteris*, BLADDER FERN, page 78
1 Blade ternate (divided into 3 more or less equal parts); indusium absent . *Gymnocarpium*
. . . . OAK FERN, page 92

Cystopteris Bernh. BLADDER FERN

CYSTOPTERIS SPECIES ARE DELICATE, monomorphic to slightly dimorphic ferns found in terrestrial habitats or growing on rocks, or in some regions as epiphytes. Species of *Cystopteris* commonly hybridize making field identification difficult. One species, *Cystopteris bulbifera,* produces vegetative reproductive structures called bulblets, which can germinate to form new plants (bulblets also rarely formed on plants of *C. laurentiana* and *C. tennesseensis,* the bulblets often mis-shapen).

KEY CHARACTERS
• Stipe grooved, hairy or smooth, vascular bundles 2, round or oblong.
• Leaf blade 1–3 pinnate-pinnatifid, ovate lance-shaped.
• Costae grooves continuous from rachis to costae, margins toothed, veins free, simple or forked, ending at margin.
• Sori round, in 1 row between midrib and margin, indusium hood-shaped, often shed at maturity, attached beneath sorus on midrib side, sporangia brown.

SIMILAR GENERA
Cystopteris and *Woodsia* are small ferns having thin-textured leaves and growing mostly on or near rock outcrops; they also frequently occur together and are often confused. Distinguishing characters are:
• *Cystopteris* has an undivided indusium, pocket-like or hood-like, attached around one side of the sorus; *Woodsia* has an indusium divided into a series of scale-like or hair-like structures, attached below sorus.
• In *Cystopteris,* stipe bases deciduous; *Woodsia* has persistent, dark stipe bases.
• In *Cystopteris,* veins reach pinnule margin; final veinlets in *Woodsia* do not extend to margin.
NAME Greek: *kystos,* bladder, and *pteris,* a fern (the ancient Greeks used *pteris* to describe all ferns); the indusium is inflated or bladder-like.

NOTE Glands in *Cystopteris* are small and may be shed during growing season, and by midsummer, glands usually absent; this is especially noticeable on *C. laurentiana* except for persistent glands on the indusia.

KEY TO *CYSTOPTERIS* | BLADDER FERN

1 Blade broadest at or near base; rachis and midribs of pinnules with small gland-tipped hairs; bulblets sometimes present on underside of rachis. 2

1 Blade broadest near middle; rachis and midribs of pinnules without glandular hairs; bulblets not present 4

2 Gland-tipped hairs usually dense on the various axes; blade triangular, widest at base and long-tapering to tip; bulblets often present on rachis of mature fronds; stipe conspicuously red in the earliest leaves; regionwide . . . . *Cystopteris bulbifera*
. . . . BULBLET BLADDER FERN, page 80

2 Gland-tipped hairs sparse; blade more abruptly tapered at tip; bulblets few, often mis-shapened ; stipe not red in earliest fronds (this pair of species can be difficult to distinguish, and both occupy similar rock habitats) 3

3 Blade usually widest at base, triangular in outline; PA only . . . . *Cystopteris tennesseensis*
. . . . TENNESSEE BLADDER FERN, page 88

3 Blade usually widest just a little above base, the outline more ovate or lance-shaped; CT, MA, PA, VT . . . . *Cystopteris laurentiana*, ST. LAWRENCE BLADDER FERN, page 84

4 Stem usually extends past the point of attachment of the current year's fronds (that is, fronds emerging slighly behind rhizome tip); rhizome with tan to golden hairs (seen by carefully removing leaf litter and soil near outermost fronds); plants typically in forest soil, less commonly on rocks; CT, NJ, NY, PA . . . . *Cystopteris protrusa*
. . . . LOWLAND BLADDER FERN, page 86

4 Stem not extending past the current fronds (fronds emerging at rhizome tip), rhizome hairs absent; usually growing on rocks. 5

5 Pinnae usually perpendicular to rachis and not curving toward tip of frond; margins with sharp teeth; plants on rock outcrops; regionwide but absent from NJ, RI . . . .
. . . . *Cystopteris fragilis*, BRITTLE BLADDER FERN, page 82

5 Pinnae often angled toward tip of frond and/or curved toward tip of frond; margins usually with low and rounded teeth; plants often on rocks, less commonly on forest floor; region-wide . . . . *Cystopteris tenuis*, UPLAND BRITTLE BLADDER FERN, page 90

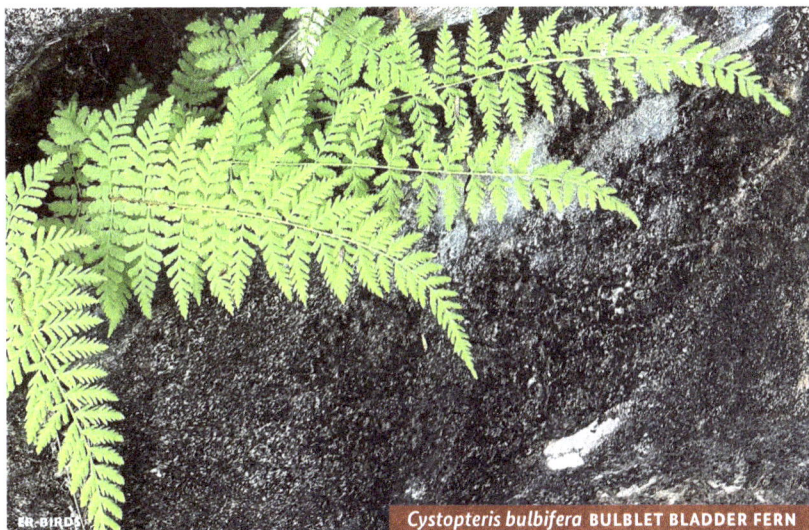

Cystopteris bulbifera BULBLET BLADDER FERN

Cystopteris bulbifera (L.) Bernh.
BULBLET BLADDER FERN

DISTRIBUTION CT | MA | ME | NH | NJ | NY | PA | RI | VT

FIELD TIPS
- loosely clumped deciduous fern; sterile and fertile fronds alike
- fronds long-tapered to tip
- pea-like bulblets on upper portion of blade
- rachis and costae with many gland-tipped hairs (small, but visible with hand lens)

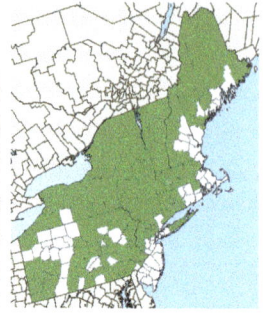

HABITAT

Moist crevices and ledges on cliffs, shaded talus slopes, hummocks in cedar swamps, usually where calcareous, rarely on soil; occasional near seeps and in moist alluvial woods.

DESCRIPTION

ROOTSTOCK Short-creeping, black, covered with old stipe bases, with a few brown scales.

FROND Deciduous, usually loosely clustered; sterile and fertile fronds alike, but early fronds smaller, sterile, less divided; to 75 cm long by 15 cm wide.

STIPE In groups of 3–8 from stem apex, grooved, maroon or reddish when young, straw-colored to green above, sparsely scaly at base, vascular bundles 2, round or oblong.

RACHIS Shiny, yellow, delicate, with gland-tipped hairs.

BLADE Narrowly triangular, widest at base, arching, 2-pinnate-pinnatifid, bearing bulblets on the rachis (or sometimes the costa) in upper 1/3 of blade; light green or yellowish-green.

PINNAE 18 to 30 pairs, perpendicular to the rachis, nearly opposite at base, becoming alternate upward, lance-shaped, basal pinnae slightly longer than next pair above; costae grooves continuous from rachis to costae, costae usually densely covered by gland-tipped hairs; margins notched; veins free, simple or forked, directed to notches between teeth.

SORI Round, in 1 row between midrib and margin; indusium on a vein, cup-shaped, beneath sorus on midrib side; sporangia brown to black.

SYNONYM *Filix bulbifera* (L.) Underw.

SIMILAR SPECIES The bulblets, common on mature fronds, distinguish *Cystopteris bulbifera* from all other *Cystopteris* except

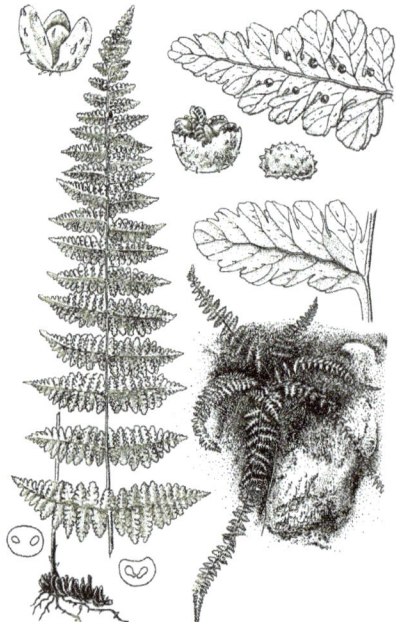

C. laurentiana and *C. tennesseensis,* where bulblets are only rarely formed.

• **Brittle fragile fern** (*Cystopteris fragilis*) but blade in that species much shorter and without glandular hairs.

NAME This North American fern was collected in Canada and sent to Linnaeus, who named it *Polypodium bulbiferum* in 1753; it was included in the new genus *Cystopteris* by Bernhardi in 1806. From Greek, *bolbos,* bulb, and Latin *fero,* to bear; referring to the bulblets.

LAURA COSTELLO

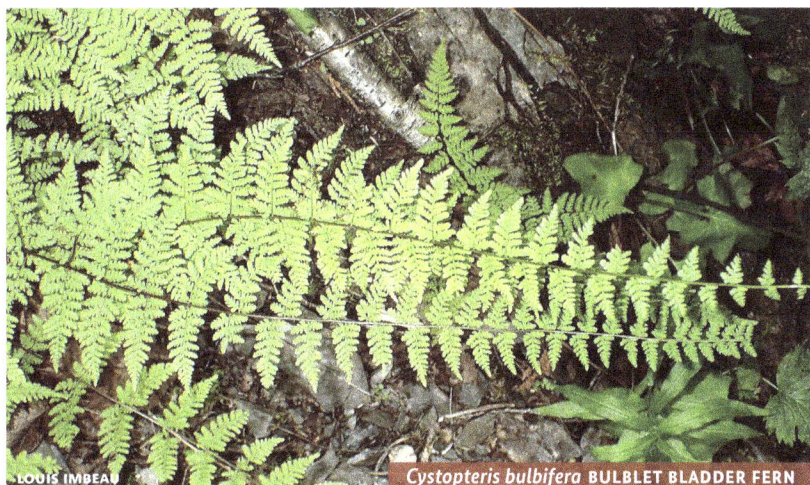

LOUIS IMBEAU   *Cystopteris bulbifera* BULBLET BLADDER FERN

Cystopteris fragilis (L.) Bernh.
BRITTLE BLADDER FERN

DISTRIBUTION CT | MA | ME | NH | NJ | NY | PA | VT

FIELD TIPS
- small, deciduous, clumped fern; sterile and fertile fronds alike; fronds may dry up during summer drought
- lower pinnae widely spaced
- veins simple or forked, usually ending at teeth
- indusium hoodlike over the sorus (but may be shed as plant matures)

HABITAT

Calcareous or acidic cliff faces, moist talus slopes, usually where partially shaded; also sometimes in moist woods and atop decaying tree stumps.

DESCRIPTION

ROOTSTOCK Short-creeping, shorter than curent year's fronds, hairy and covered with old stipe bases; scales few, tan to light brown, lance-shaped.

FROND Deciduous, clustered; sterile and fertile fronds alike, to 40 cm long.

STIPE 3- to 8-clustered at stem apex, grooved, dark at base, straw-colored to green above, sparsely scaly, vascular bundles 2, round or oblong.

RACHIS Smooth, without scales.

BLADE Ovate lance-shaped, 2-pinnate-pinnatifid, smooth.

PINNAE 9 to 15 pairs; opposite or nearly so, lance-shaped, widest just below the middle, perpendicular to the rachis, lowest pinnae pairs bending forward and down; costae grooves continuous from rachis to costae; margins variably lobed, toothed to smooth; veins free, simple or forked, usually running to teeth.

SORI Round, in 1 row between midrib and margin; indusium forming a hood over the sori, but shriveling with maturity, attached beneath sori on midrib side; sporangia brown to black.

SYNONYMS *Cystopteris dickieana* Sim, *Filix fragilis* (L.) Underw.

SIMILAR SPECIES
- **Lowland bladder fern** (*Cystopteris protrusa*) but blade of *C. protrusa* wider (less than 2.5 times longer than wide); blade of *C. fragilis* narrower (more than 2.5 times longer than wide).

• **Bulblet bladder fern** (*Cystopteris bulbifera*), but *C. fragilis* never producing bulblets.
• **Upland brittle bladder fern** (*Cystopteris tenuis*), but blade of this species larger and pinnae angled toward frond tip (perpendicular in fragile fern).
• **Blunt-lobe cliff fern** (*Woodsia obtusa*), which has a few scales on rachis, these are absent in *Cystopteris*.

NOTE In earlier classifications, *Cystopteris fragilis* included *C. laurentiana, C. tennesseensis, C. protrusa,* and *C. tenuis* as varieties.

NAME Named *Polypodium fragile* by Linnaeus in 1753, and referred to a new genus, *Cystopteris,* by Bernhardi in 1806. From Latin, *fragilis,* easily broken; referring to the fragile stipes that are easily broken when bent.

DAVE RICHARDSON

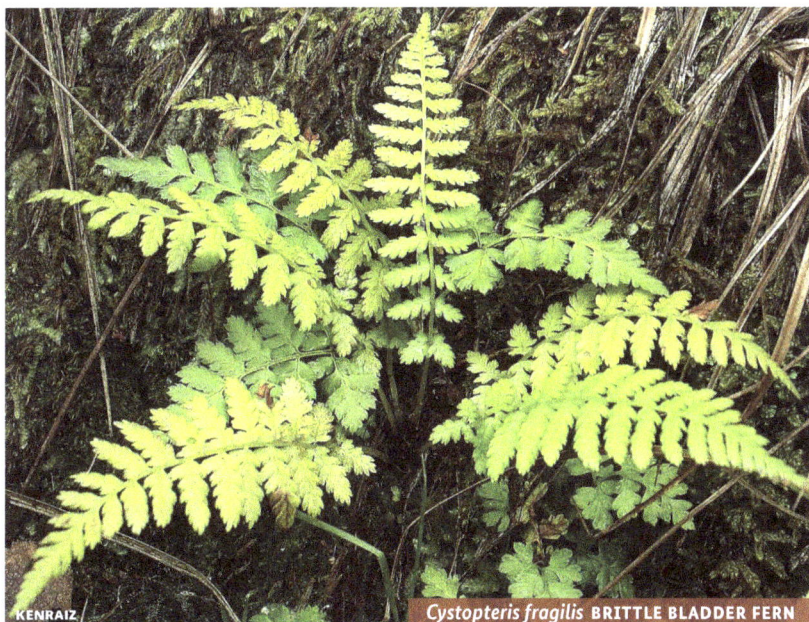

KENRAIZ   *Cystopteris fragilis* BRITTLE BLADDER FERN

Cystopteris laurentiana (Weatherby) Blasdell
ST. LAWRENCE BLADDER FERN

DISTRIBUTION CT | MA | PA | VT

FIELD TIPS
- hybrid between *Cystopteris bulbifera* and *C. fragilis*
- blade ovate in outline, with sparse gland-tipped hairs
- sometimes with a few small, scaly, misshapen bulblets
- moist calcareous rock habitats

HABITAT
Cracks and ledges on calcareous rock outcrops, usually in partial or full shade.

DESCRIPTION

ROOTSTOCK Short-creeping, covered with old stipe bases; scales tan to light brown, lance-shaped.

FROND Deciduous, loosely clustered; sterile and fertile fronds alike, nearly all fronds fertile; to 50 cm long by 12 cm wide.

STIPE Clustered at tip of rootstock, grooved, brown at base to straw-colored above, sparsely scaly at base; vascular bundles 2, round or oblong.

BLADE Ovate, usually widest above base, rarely with bulblets on the rachis, 2-pinnate-pinnatifid, light green or yellowish-green, sparse gland-tipped hairs present on rachis and costa.

PINNAE 12 to 16 pairs; perpendicular to the rachis, ± opposite, lance-shaped; costae grooves above continuous from rachis to costae; margins serrate; veins free, simple or forked, running to teeth and notches.

SORI round, in 1 row between midrib and margin; indusium cup-shaped, beneath sorus on midrib side, on a vein; sporangia brown to black.

SYNONYMS *Cystopteris fragilis* var. *huteri* (Hausman) Luerss., *Cystopteris fragilis* var. *laurentiana* Weath.

SIMILAR SPECIES
- Similar to **bulblet bladder fern** (*Cystopteris bulbifera*), but plants smaller, blade widest above base, with only sparse glandular hairs, bulblets rare or absent, stipe brown, not maroon or reddish, and veins running to tips of teeth and notches.
- *Cystopteris laurentiana* larger than **brittle bladder fern** (*C. fragilis*), with more pinnae, sometimes with a few small, reddish, misshapen bulblets; and veins sometimes running to notches; in *C. fragilis* veins mostly to teeth-tips only.

NOTE Believed to have originated as a hybrid between *Cystopteris bulbifera* and *C. fragilis*.

NAME Laurentiana, for the St. Lawrence River and surrounding region.

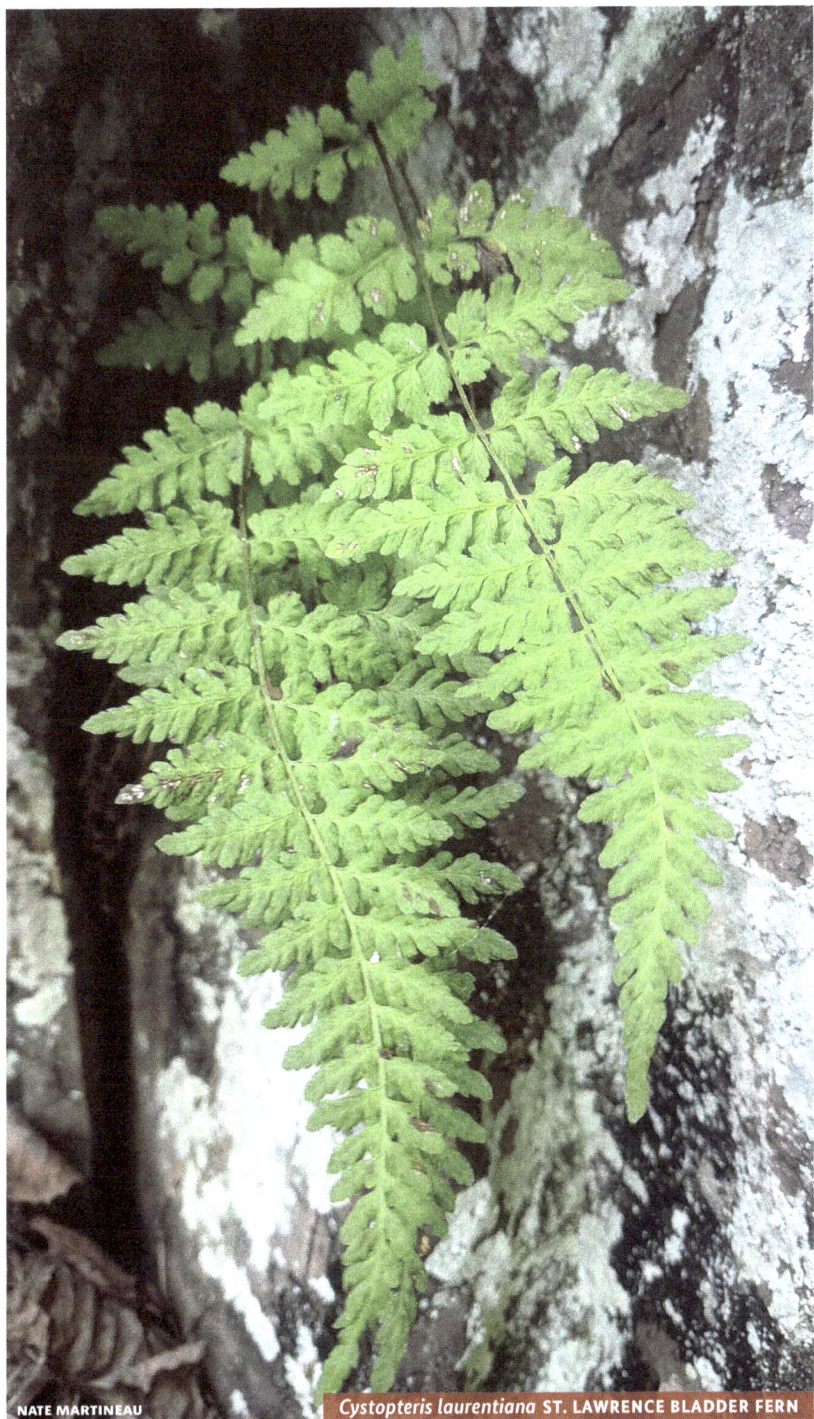

NATE MARTINEAU

Cystopteris laurentiana ST. LAWRENCE BLADDER FERN

Cystopteris protrusa (Weatherby) Blasdell
LOWLAND BLADDER FERN

DISTRIBUTION CT | NJ | NY | PA

FIELD TIPS
- deciduous fern of moist woods
- usually in soil rather than rock
- blade lance-shaped; pinnules of lower pinnae distinctly stalked
- rhizome tip extends past emergent fronds

HABITAT

Soil of rich, moist forests and alluvial woods; rarely on rock; can form colonies from its creeping rhizomes.

DESCRIPTION

ROOTSTOCK Creeping, extending about 2–3 cm beyond current year's fronds, covered with old stipe bases and with tan to golden hairs, and with light brown scales, these mostly near rhizome tip.

FROND Deciduous, loosely clustered; sterile and fertile fronds alike, but early fronds smaller, sterile, less divided; 40 cm long by 10 cm wide.

STIPE Rising 1–several cm behind the rhizome tip, grooved, dark at base, straw-colored to green upwards; scales at base; vascular bundles 2, round or oblong.

RACHIS Smooth.

BLADE Ovate lance-shaped, 2-pinnate-pinnatifid, smooth.

PINNAE 6 to 12 pairs; lance-shaped, widest just below the middle, lowest pinnae pair usually bending forward and down; costae grooves continuous from rachis to costae; veins free, simple or forked, usually ending at tooth.

SORI Round, in 1 row between midrib and margin; indusium ovate, forming hood over the sorus, but shriveling with maturity, beneath sorus on midrib side; sporangia brown to black.

SYNONYMS *Cystopteris fragilis* var. *protrusa* Weath.

CONSERVATION STATUS *endangered* New York.

SIMILAR SPECIES Distinguished from other bladder ferns (*Cystopteris*) by growing in soil rather than on rocks, and by the protruding, golden-haired rhizome (rhizome tip can be felt with fingertips just beyond outermost emerging fronds).

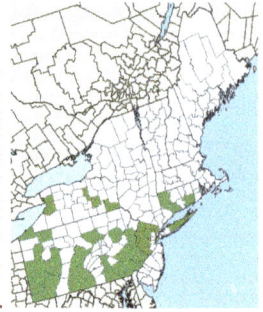

NOTE Separated from *Cystopteris fragilis* in 1935, when C. A. Weatherby named it variety *protrusa;* now considered a distinct species.

NATE MARTINEAU

Cystopteris protrusa **LOWLAND BLADDER FERN**

Cystopteris tennesseensis Shaver
TENNESSEE BLADDER FERN

DISTRIBUTION PA

FIELD TIPS
- deciduous fern of rock cliffs
- hybrid between *Cystopteris bulbifera* and *C. protrusa*, appearance intermediate between the two parents
- bulblets present or not, often misshapen, dark, scaly

HABITAT

Cracks and ledges of limestone and sandstone rocks, with best growth in calcium-rich habitats; rarely in soil.

DESCRIPTION

ROOTSTOCK Creeping, not cordlike, with many old stipe bases; not hairy, scales lance-shaped, light brown

FRONDS Crowded near tip of rootstock, sterile and fertile stems alike; to 45 cm long, nearly all fronds with sori.

STIPE Mostly dark brown at base, gradually becoming straw-colored upwards, sparsely scaly at base.

RACHIS With or without bulblets (if present usually misshapen); axils of pinnae sometimes with gland-tipped hairs.

BLADE Triangular in outline, short-tapered to tip, 2-pinnate-pinnatifid, usually widest at or near base.

PINNAE Usually perpendicular to rachis, not curving toward blade tip; margins serrate; veins running to sinuses and teeth.

SORI Round; indusium cup-shaped, with scattered gland-tipped hairs.

SYNONYMS *Cystopteris fragilis* var. *simulans* (Weath.) McGregor, *Cystopteris fragilis* var. *tennesseensis* (Shaver) McGregor

SIMILAR SPECIES Similar to bulblet bladder fern (*Cystopteris bulbifera*) but differs in having only a few scattered glandular hairs, in having fewer bulblets, the bulblets misshapen, scaly and dark (vs. smooth and green in *C. bulbifera*), the sori larger (2–5 mm wide vs. 1–2 mm in *C. bulbifera*) and in lacking red stipes in young plants. Also, in *C. tennesseensis*, veins run to sinuses and teeth whereas in *C. bulbifera*, veins run to sinuses. Also similar to our other species of *Cystopteris*, see key, page 79.

NOTE Believed to have originated from an ancient cross between *Cystopteris bulbifera* and *C. protrusa*. *C. tennesseensis* may also hybridize with *C. tenuis* to form **Cystopteris × wagneri**. *C. tennesseensis* may grow to-

gether with other members of the genus in mixed populations.

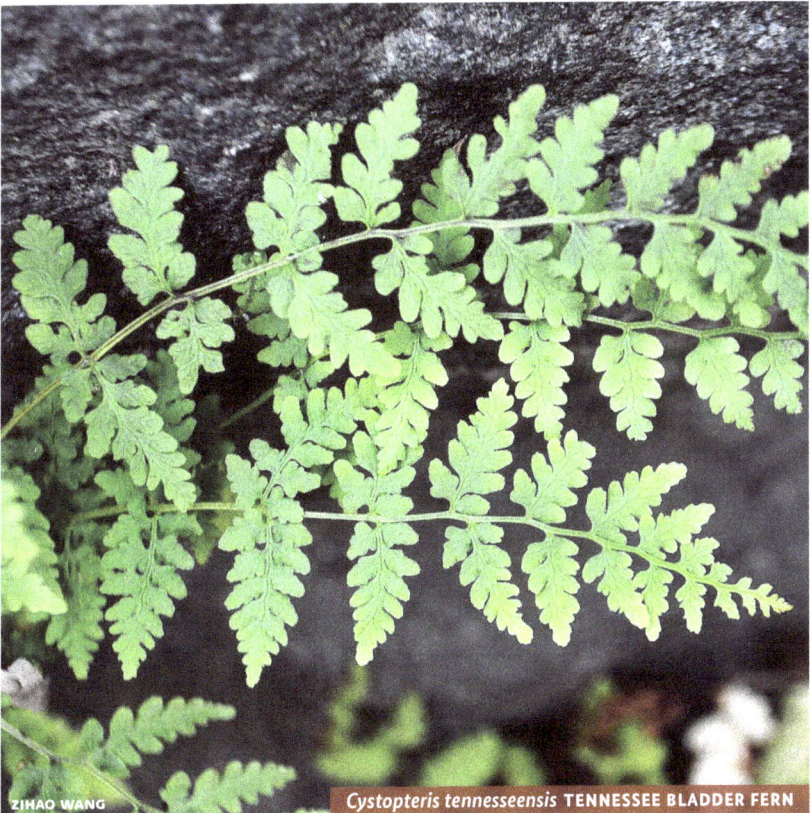

Cystopteris tennesseensis TENNESSEE BLADDER FERN

Cystopteris tenuis (Michx.) Desv.
UPLAND BRITTLE BLADDER FERN

DISTRIBUTION CT|MA|ME|NH|NJ|NY|PA|RI|VT
FIELD TIPS
• deciduous, bright green, loosely clumped fern
• usually found on rock but also in soil

HABITAT
Shaded, mossy rock outcrops, sandstone cliffs; less commonly in soil in moist forests; forming small patches but not extensive colonies.

DESCRIPTION
ROOTSTOCK Short-creeping, covered with old stipe bases; scales tan to light brown, lance-shaped.

FROND Deciduous, loosely clustered; sterile and fertile fronds alike; to 40 cm long.

STIPE Grooved, dark brown at base, straw-colored above; sparsely scaly at base; vascular bundles 2, round or oblong.

BLADE Ovate lance-shaped, 2-pinnate-pinnatifid, smooth.

PINNAE ca. 12 pairs, ± opposite, lance-shaped, widest just below the middle, angled upward from rachis and curving towards tip; costae grooves continuous from rachis to costae; margins lobed, with rounded teeth; veins free, simple or forked, directed into teeth and notches.

SORI Round, in 1 row between midrib and margin; indusium ovate to lance-shaped, forming a hood over the sorus, but shriveling with maturity, attached beneath sorus on midrib side; sporangia brown to black.

SYNONYM *Cystopteris fragilis* var. *mackayi* G. Lawson

SIMILAR SPECIES *Cystopteris tenuis* previously considered a variety of **brittle bladder fern** (*C. fragilis*) and is similar in appearance to that species and **lowland bladder fern** (*C. protrusa*); the pinnae of *Cystopteris tenuis* tend to be more curved in shape and angled toward tip of frond, and margins with rounded teeth rather than sharp teeth. Also differs from *Cystopteris protrusa* in having fronds clustered at end of a short, rather than long-creeping rhizome.

NAME Named by Linnaeus *Polypodium fragile* in 1753 (original name for *Cystopteris fragilis*), and referred to new genus

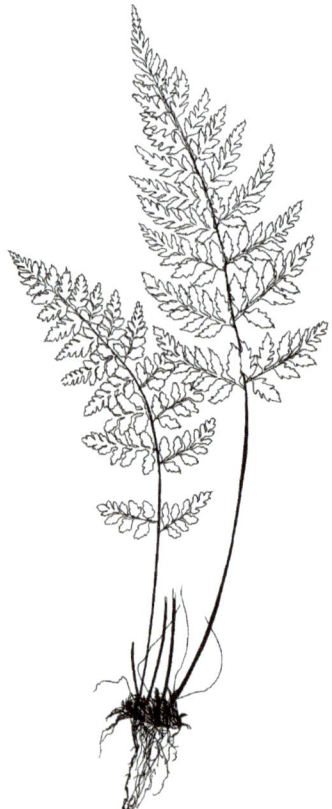

Cystopteris by Bernhardi in 1806. The varietal name *mack-ayi* was applied by Lawson in 1889.

AARON GUNNAR

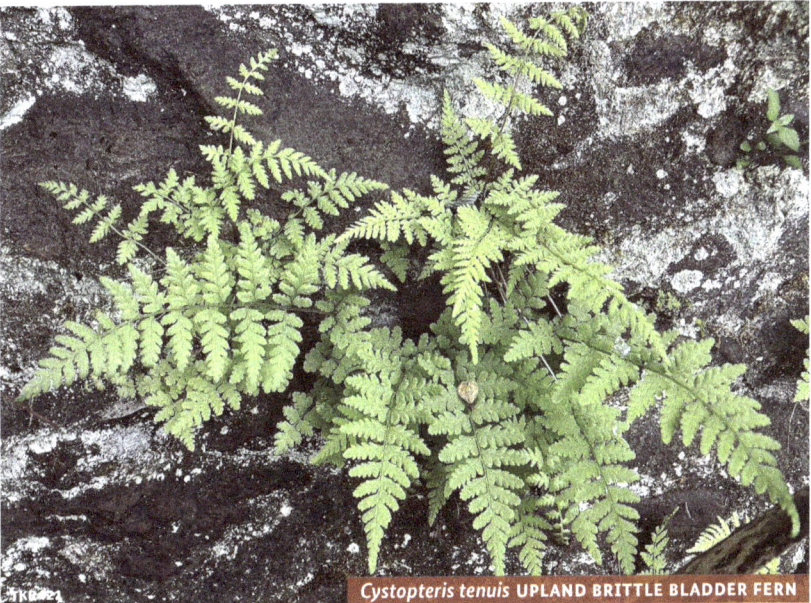

Cystopteris tenuis UPLAND BRITTLE BLADDER FERN

Gymnocarpium Newman OAK FERN

GYMNOCARPIUM ARE SMALL TO MEDIUM, delicate ferns of cool, shaded places, either in soil or on rock. *Gymnocarpium* includes 7 species, with 5 occurring in temperate regions of North America; three in our flora, but only *G. dryopteris* is common.

KEY CHARACTERS
- Plants forming small colonies from long-creeping, smooth, dark rhizomes.
- Frond deciduous, sterile and fertile fronds alike.
- Stipe grooved, smooth; vascular bundles 2, oblong.
- Leaf blade nearly horizontal, pinnatifid to 3-pinnate-pinnatifid, triangular to pentagonal in outline.
- Pinnules on lower side of pinnae longer than those on upper side, costae grooved above, not continuous from rachis to costae, segments entire or crenate, veins free, simple or forked.
- Sori small, round, in 1 row between midrib and margin, indusium absent, sporangia brownish.

NAME From the Greek, *gymnos,* naked, *karpos,* fruit, because there are no indusia covering the sori.

ADDITIONAL NORTHEASTERN SPECIES
- **Appalachian Oak Fern (*Gymnocarpium appalachianum* Pryer); rare in Pennsylvania (state endangered); endemic to the central and southern Appalachian Mountains; found in moist, rocky wood, often at higher elevations; very similar to the common *G. dryopteris* (see key).
- **Asian Oak Fern [*Gymnocarpium continentale* (Petrov.) Pojark. {synonym: *Gymnocarpium jessoense* subsp. *parvulum* Sarvela); reported for Massachusetts, historical record from southern Vermont, becoming more common northward; usually on or near non-calcareous rock or talus, in shade or partial sun. Fronds deciduous, arising singly from the creeping rhizomes; fronds usually less than 40 cm long; blade broadly triangular, 2-pinnate-pinnatifid and ternate (divided into three roughly equal branches at the base); sori round, located on underside of blade; indusium absent (see key).

NOTE In *Gymnocarpium continentale,* underside of rachis and blade clearly glandular and upperside smooth; lowest pinna pair curved toward tip of blade; *G. dryopteris* smooth (without hairs or glands) on both sides.

Gymnocarpium appalachianum

JASON GRANT

Gymnocarpium continale ASIAN OAK FERN

Gymnocarpium dryopteris (L.) Newman
NORTHERN OAK FERN

DISTRIBUTION CT | MA | ME | NH | NJ | NY | PA | RI | VT

FIELD TIPS
- delicate deciduous fern of moist woods, sometimes on rock
- frond 3-parted, bright green, blades triangular, tilted to near horizontal position
- indusium absent

HABITAT

Moist woods of all types, sometimes on cliffs and boulders.

DESCRIPTION

ROOTSTOCK Long-creeping and branching, slender, dark brown to black, scaly.

CROZIER Delicate, green, 3-parted, one for each developing pinna; produced all summer.

FROND Deciduous, singly from the rhizome; sterile and fertile fronds alike; to 40 cm long.

STIPE Grooved, wiry, straw-colored, darker, purplish at the base, scaly at base, vascular bundles 2, oblong at stipe base, becoming round above.

RACHIS Green to dark purplish-green, delicate, sometimes glandular, bending so that blade more or less horizontal.

BLADE Broadly triangular, appearing 3-parted in outline, 2-pinnate-pinnatifid, vivid green; uppermost pinna arched to more or less horizontal position; lower 2 pinna nearly parallel to ground; smooth, or underside with a few glandular hairs.

PINNAE 6 to 10 pairs; the lowest pair, exactly opposite, attached at a swollen junction, each similar to the remainder of the blade, only slightly smaller; the lower, basal pinnule of the basal pinna most divided; costae grooved, continuous from rachis to costae; margins entire to crenate; veins free, simple or forked.

SORI Round, in rows near the margin, often merging at maturity; indusium absent; sporangia brownish.

SYNONYMS *Dryopteris dryopteris* (L.) Britton, *Dryopteris linnaeana* C. Chr., *Phegopteris dryopteris* (L.) Fée, *Thelypteris dryopteris* (L.) Slosson

CONSERVATION STATUS *threatened* Rhode Island.

SIMILAR SPECIES In the less common **Asian oak fern** (*Gymnocarpium continentale*), the rachis and blade underside are clearly

glandular (upperside smooth); *G. dryopteris* lacks glands.

NOTE Believed to have originated as hybrid between *Gymnocarpium appalachianum* and *G. disjunctum*.

NAME Known to Linnaeus in Europe, and named by him *Polypodium dryopteris* in 1753; included in *Phegopteris* by Fée nearly 100 years later. From Greek, *drys,* oak plus *pteron,* a wing, describing shape of the pinnae; *pteris* was used by the ancient Greeks for all ferns.

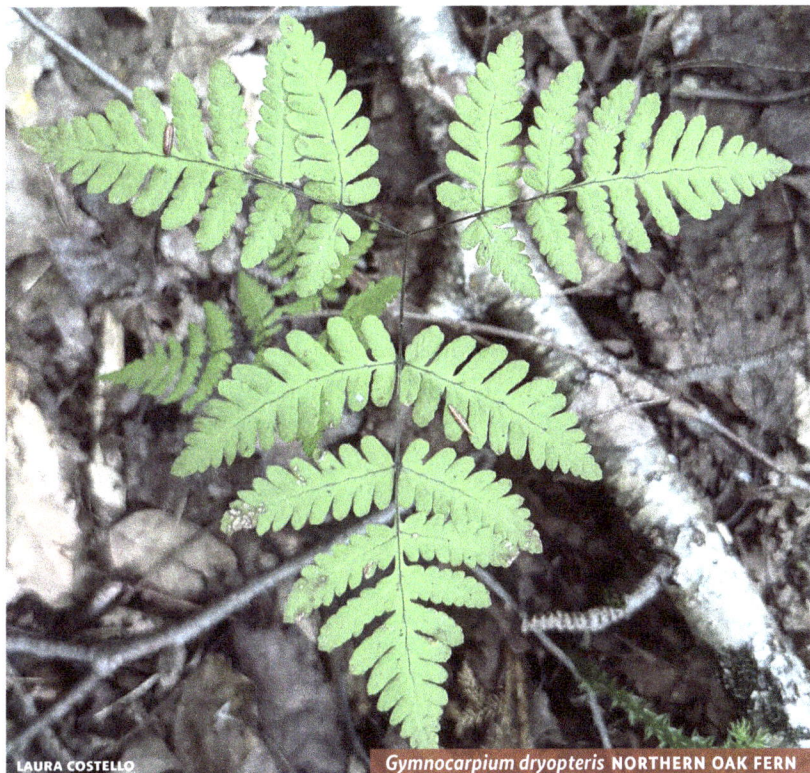

LAURA COSTELLO

Gymnocarpium dryopteris **NORTHERN OAK FERN**

DENNSTAEDTIACEAE *Bracken fern family*

LARGE, DECIDUOUS, COLONY-FORMING FERNS. Sori on blade margins, either in round separate clusters covered by a cup-shaped indusium (*Dennstaedtia*); or in *Pteridium,* more or less continuous along margin of frond and covered by recurved border of the leaf segment (forming an outer false indusium). Worldwide, 10 genera and an estimated 265 species, mostly of tropical regions; 2 genera in our flora.

KEY TO DENNSTAEDTIACEAE | BRACKEN FERN FAMILY

1 Leaf blades elongate in outline, at least 4x as long as broad, not leathery; sori globular, separate; common, regionwide . . . . *Dennstaedtia punctilobula*
. . . . HAY-SCENTED FERN, page 96

1 Leaf blades broadly triangular in outline, about as broad as long, somewhat leathery; sori linear, continuous along margins as a marginal band; common, regionwide. . . . .
. . . . *Pteridium aquilinum*, NORTHERN BRACKEN FERN, page 98

Dennstaedtia Bernh. HAY-SCENTED FERN

LARGE DECIDUOUS FERNS. Worldwide, about 70 species, mostly in the tropics; one species in our region. Sometimes weedy in northeastern USA where it may form large, crowded colonies.

KEY CHARACTERS
• Colony-forming from slender creeping rhizomes.
• Sterile and fertile fronds alike; blade covered with small, white, gland-tipped hairs, these hay-scented (especially as drying).
• Indusia cup-shaped; sori at tip of vein on margin of pinnule lobes.

NAME Named for German botanist, August Wilhelm Dennstaedt (1776–1826).

Dennstaedtia punctilobula (Michx.) T. Moore
HAY-SCENTED FERN

DISTRIBUTION CT | MA | ME | NH | NJ | NY | PA | RI | VT
FIELD TIPS
• large deciduous fern
• fronds single from the rhizome, may form dense colonies
• fronds with glandular hairs on upper and lower surface, margins round-toothed
• indusium cup-shaped, sorus at end of vein on blade margin

HABITAT
Partially shaded to sunny woods, roadsides, fields, clearings, rocky slopes.

DESCRIPTION
ROOTSTOCK Long-creeping, slender, with dark, red-brown, jointed hairs near new growth; scales few or absent.
CROZIER Covered with silver-white, gland-tipped hairs.

FROND Deciduous, arising singly along the rhizome and forming colonies; sterile and fertile fronds alike.

STIPE Shorter than blade, straw-colored to brown, darker to nearly black at base, grooved above, glandular hairy, vascular bundles 1, arranged in a U-shape.

RACHIS Pale to straw-colored, slender, hairy.

BLADE Lance-shaped, lacy, papery, 2-pinnate-pinnatifid (sometimes less or more divided), yellow-green to pale green, with silver-gray, jointed hairs on both surfaces and gland-tipped hairs containing a fragrant wax below.

PINNAE Broadest at base, segments ovate to lance-shaped, margins with rounded teeth, veins free, pinnately branched.

SORI Globose to almost cylindric, marginal at vein tips; indusium a circular cup.

SIMILAR SPECIES

• **Lady fern** (*Athyrium* spp.) is clumped, not spreading by rhizomes; stipe with dark brown scales, not hairs.

• Blade of **New York fern** (*Amauropelta noveboracensis*) gradually tapered to base, lowest pinnae very small.

NAME Michaux discovered this fern in Canada and named it *Nephrodium punctilobulum* in 1803. It was later mistakenly transferred to the tropical genus *Dicksonia*. In 1857, however, Moore showed it to belong in Bernhardi's genus *Dennstaedtia*. From Latin, *punctum*, small spot, *lobulus*, small lobe; referring to the sori appearing as small spots on the pinnule lobes.

Dennstaedtia punctilobula **HAY-SCENTED FERN**

ALINA MARTIN

Pteridium Gleditsch ex Scop. **BRACKEN FERN**

Coarse deciduous ferns, spreading by rhizomes and sometimes covering large areas. Worldwide, the number of species is uncertain but likely number less than 20; one species in the region.

KEY CHARACTERS
- Large coarse, deciduous ferns, sterile and fertile fronds alike, with long, creeping, hairy rhizomes.
- Blade generally triangular-shaped, mostly 3-pinnate.
- Sori continuous, along the margin, indusium double, the outer false and formed by the recurved border of the leaf segment, the inner true indusium tiny or absent.

NOTE *Pteridium* is the world's most widely distributed genus of fern, sometimes becoming a serious weed and very difficult to eradicate due to the deeply seated rhizomes. Bracken ferns also release allelopathic chemicals, which in combination with its shady canopy and thick litter, inhibit other plant species from establishing.

NAME From Greek, *pteridion,* a small fern. Bracken is of Old Norse origin, related to the Swedish word *bräken,* meaning fern.

Pteridium aquilinum (L.) Kuhn
NORTHERN BRACKEN FERN

DISTRIBUTION CT | MA | ME | NH | NJ | NY | PA | RI | VT
FIELD TIPS
- large deciduous fern, very common
- fronds in rows from rhizomes, often forming large colonies
- blade broadly triangular in outline, held almost parallel to the ground

HABITAT

Dry, acidic, sandy woods, clearings, abandoned fields, waste places, burned areas, in sun to partial shade; often forming large colonies from extensive and long-creeping rhizomes; becoming a serious, difficult to eradicate weed in some places.

DESCRIPTION

ROOTSTOCK Long-creeping and branching, cordlike, to 2 cm wide, deep-seated (30 cm or more deep); scales absent.

CROZIER With 3 sections, uncoiling like opening of an eagle's claw, covered with silvery gray hairs.

FROND Deciduous, scattered on the rhizome; sterile and fertile fronds alike, to about 1 m tall.

STIPE Variable in length, somewhat woody, dark purple-brown at base, straw-colored above; vascular bundles numerous (often more than 10), of various sizes and shapes.

RACHIS Green, grooved, slightly hairy.

BLADE Large, broadly triangular, 2- to 4-pinnate, waxy.

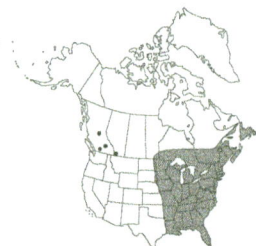

PINNAE Lowest very large; pinnules variable; costae rachis and costae grooved above, veins free, forked, except for a marginal strand.

SORI Continuous, along the margin; indusium double, the outer false, reflexed, the inner distinct or absent.

NOTE Bracken fern fiddleheads contain the carcinogenic terpene ptaquiloside and should not be eaten.

NAME Linnaeus named the European bracken *Pteris aquilina* in 1753. Seven years later Scopoli placed it in genus *Pteridium,* on the basis of the special indusium. *Aquilinum,* from Latin *aquilus,* eagle-like; perhaps referring to talon-like shape of the unfurling crozier.

Pteridium aquilinum **NORTHERN BRACKEN FERN**

DIPLAZIOPSIDACEAE *Hanging Fern Family*

Homalosorus Small ex Pichi Sermolli GLADE FERN

ONCE PLACED IN THE FAMILY ATHYRIACEAE due to its linear, often doubled sori, our single genus in this family, *Homalosaurus,* is now (as of 2016, PPG I) placed in the small family Diplaziopsidaceae (other sources place the genus in subfamily Diplaziopsidoideae of a more broadly defined Aspleniaceae). Worldwide, two genera and four species are defined, with *Homalosaurus pycnocarpos* found in eastern North America (the other genus with its three species are found in east Asia).

KEY CHARACTERS
- large deciduous fern, fronds clumped or solitary.
- blade 1-pinnate; pinna margins wavy, not sharp-toothed.
- sori in herringbone pattern, silvery green when young.
- Indusium elongate, may appear raised due to expanding sporangia beneath.

SIMILAR GENERA Similar to lady fern (*Athyrium*), but sori in
Homalosorus never hook over the veins, and are sometimes paired back-to-back; also, grooves in the rachis are
U-shaped vs. V-shaped in *Athyrium.* Glade fern differs
from silvery spleenwort (*Deparia*) in having the grooves
continuous from costa to rachis.

NAME Family name from Greek, *diplazios,* double, *plasion,*
oblong; referring to a 2-valved indusium.

Homalosorus pycnocarpos (Spreng.) Pichi Sermolli
GLADE FERN

DISTRIBUTION CT | MA | NH | NJ | NY | PA | VT
FIELD TIPS
- large deciduous fern, fronds clumped or solitary
- blade 1-pinnate; pinna margins wavy, not sharp-toothed
- sori in herringbone pattern, silvery green when young

HABITAT
Moist, open to partially shaded woods, meadows and slopes;
soils rich and circumneutral.

DESCRIPTION
ROOTSTOCK Short-creeping, stipes 1–2 cm apart; scales
brown or tan, broadly lance-shaped.

FROND Deciduous, solitary or clustered; sterile and fertile
fronds somewhat different; fertile fronds taller, erect; sterile fronds arching; 100 cm long by 20 cm wide.

STIPE About 1/2 as long as blade (stipe of fertile frond
longer and stiffer than those of sterile fronds), hairy (at
least when young), reddish-brown and scaly at base, green
upwards, deeply grooved above; vascular bundles 2, uniting to a V-shape above.

RACHIS Green to straw colored, underside slightly hairy.

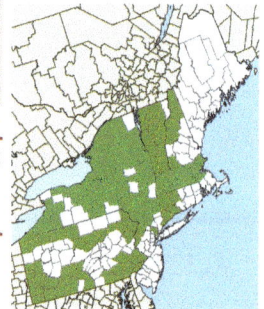

BLADE Oblong-lance-shaped, ± narrowed to base, 1-pinnate.

PINNAE 20 to 40 pairs, narrow, to 1 cm wide, tapered to long sharp tip, base rounded to heart-shaped, lowest pinnae with very short stalk; costae grooved above, continuous from rachis to costae; smooth; margins entire to crenate; veins free, forking. Sterile pinnae often somewhat twisted, not stiff; fertile pinnae narrowed, straight and stiff.

SORI Linear or slightly curved , running from midvein halfway or almost to margin in a herringbone pattern; indusium linear, persistent, translucent, laterally attached; sporangia dark brown when mature.

SYNONYMS *Asplenium pycnocarpon* Spreng., *Athyrium pycnocarpon* (Spreng.) Tidestr., *Diplazium pycnocarpon* (Spreng.) Broun

CONSERVATION STATUS *endangered* New Hampshire, New Jersey.

SIMILAR SPECIES The 1-pinnate leaf blade and linear sori in a herringbone pattern are distinctive.

NAME Discovered by Michaux "on the banks of the Ohio River," and named by him *Asplenium angustifolium* in 1803. However, this name was already in use for a tropical fern, and Sprengel proposed *Asplenium pycnocarpum* in 1804; Tidestrom proposed *Athyrium pycnocarpum* in 1906.

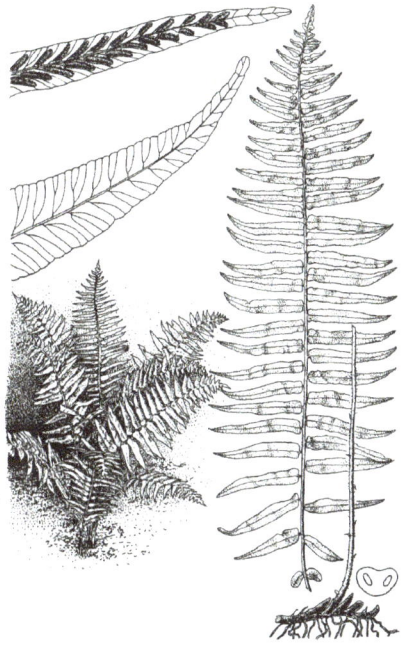

R. A. NONENMACHER

Homalosorus pycnocarpos **GLADE FERN**

DRYOPTERIDACEAE *Wood fern family*

SMALL TO LARGE FERNS growing in mineral or organic soils, or in rock crevices; plants usually clumped, sterile and fertile fronds generally more or less alike. Worldwide 26 genera and an estimated 2115 species, especially in tropical regions and in forests; 3 genera in our flora.

KEY CHARACTERS

- Usually clumped ferns of wet to dry habitats.
- Blade 1 to 4-pinnate, often with scales, hair-like scales, and/or hairs (except clear, needle-like hairs are generally absent, and are typical in Thelypteridaceae); ultimate leaf segments (smallest subdivision of blade) not entire; veins free.
- Sori round, in most species on veins or vein tips of lower blade surface (usually not marginal); indusia peltate (umbrella-like) or kidney-shaped.

NOTE The family previously included a number of other genera: *Athyrium, Cystopteris, Deparia, Homalosorus, Matteuccia, Onoclea* and *Woodsia.* These are now separated into several new families as treated in this Flora.

NAME From Greek, *drys,* oak, and *pteris,* fern.

ADDITIONAL NORTHEASTERN SPECIES

- Japanese net-vein holly fern [*Cyrtomium falcatum* (L. f.) K. Presl]; evergreen clumped fern, introduced from Asia and reported from MA and NJ (but presence and persistence not verified). *Cyrtomium* shares the peltate indusium with *Polystichum* and also sometimes has an upward ear or auricle at base of pinnae; differs in having a terminal pinna similar to the lateral ones; it shares the grooves continuous from rachis to costae with *Dryopteris;* differs from both genera in having netlike rather than free veins.

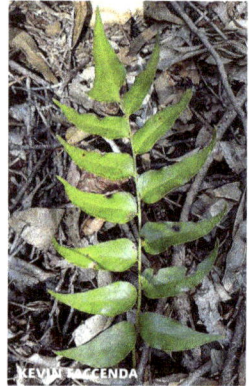

Cyrtomium falcatum
**JAPANESE NET-VEIN
HOLLY FERN**

KEY TO DRYOPTERIDACEAE | WOOD FERN FAMILY

1   Fronds 1-pinnate-pinnatifid to more divided, the pinnae pinnatifid or themselves fully divided, lacking a prominent basal lobe, light green to dark green, herbaceous to nearly leathery; indusia kidney-shaped . . . . *Dryopteris*, WOOD FERN, page 102

1   Fronds 1-pinnate, the pinnae toothed and each with a slight to prominent lobe near the base on the side towards the leaf tip, dark green, leathery or nearly so; indusia peltate (umbrella-like). 2

2   Uncommon introduced fern, in northeastern region reported from MA and NJ; pinnae 4–10 pairs per frond, veins rejoining to form a netlike pattern . . . . *Cyrtomium falcatum*
. . . . JAPANESE NET-VEIN HOLLY FERN, page 102

2   Native, mostly widespread species; pinnae many, 25–50 pairs on larger leaves; veins branching, free and not rejoining in a netlike pattern . . . . *Polystichum*
. . . . HOLLY FERN, SWORD FERN, page 126

Dryopteris Adans. WOOD FERN

MOSTLY MEDIUM TO LARGE FERNS, with scaly, stout stipes and variably dissected blades. Worldwide, about 400 species of *Dryopteris,* most commonly in temperate regions of Asia, and 10 species (plus a number of hybrids) are known from the region. Wood ferns, as the name implies, are often found in forests, but our species are found in habitats ranging from wetlands to moist or dry forests to talus and rock outcrops. Some species commonly hybridize, complicating field identification; the region's reported hybrids are listed below. *Dryopteris* similar in many respects to *Polystichum,* the latter differing in the peltate (umbrella-shaped) indusium and lack of continuity in grooves between rachis and costa.

KEY CHARACTERS
• Fronds range from 1-pinnate-pinnatifid to 3-pinnate-pinnatifid; no simple or pinnatifid *Dryopteris* are known from our region.
• Stipes scaly, without hairs (except in *D. intermedia*), the vascular bundles in a C-shape.
• Upperside of stipe, rachis, and costa with a continuous groove.
• Indusium kidney-shaped over a round sorus.

NAME From Greek, *drys,* oak, referring to the many species that are found in oak woodlands, and *pteron,* a wing, which describes the shape of the pinnae (*pteris* was used by the ancient Greeks for all ferns). The dryads of Greek mythology were the nymphs that inhabited oaks. In Britain and Europe, *Dryopteris* known as shield ferns or buckler ferns, a buckler being a small, round shield, in reference to the shape of the indusium. Previously, wood ferns were classified as *Aspidium* or *Lastrea.*

DRYOPTERIS HYBRIDS A number of *Dryopteris* hybrids occur in the northeastern region, and reported taxa are listed below. Hybrids may be recognized by an appearance intermediate between the parent species, and the presence of abortive spores.

Dryopteris marginalis
MARGINAL WOOD FERN

• **Dryopteris × benedictii** Wherry: *D. carthusiana* × *D. clintoniana*
• **Dryopteris × boottii** (Tuckerman) Underwood: *D. cristata* × *D. intermedia*
• **Dryopteris × burgessii** Boivin: *D. clintoniana* × *D. marginalis*
• **Dryopteris × correllii** W.H. Wagner: *D. carthusiana* × *D. goldieana*
• **Dryopteris × dowellii** (Farw.) Wherry: *D. clintoniana* × *D. intermedia*
• **Dryopteris × leedsii** Wherry: *D. celsa* × *D. marginalis*
• **Dryopteris × mickelii** J.H. Peck: *D. clintoniana* × *D. goldieana*
• **Dryopteris × montgomeryi** Fraser-Jenk. & Widen: *D. filix-mas* × *D. marginalis*
• **Dryopteris × neowherry** W.H. Wagner: *D. goldieana* × *D. marginalis*
• **Dryopteris × pittsfordensis** Slosson: *D. carthusiana* × *D. marginalis*
• **Dryopteris × separabilis** Small (pro sp.) (Farw.) Wherry: *D. celsa* × *D. intermedia*

- **Dryopteris** × **slossoniae** Wherry ex Lellinger: *D. cristata* × *D. marginalis*
- **Dryopteris** × **triploidea** Wherry: *D. carthusiana* × *D. intermedia*.
- **Dryopteris** × **uliginosa** (A. Braun ex Dowell) Druce: *D. carthusiana* × *D. cristata*. Usually in swamps and wet woods.

KEY TO *DRYOPTERIS* | WOOD FERN

1 Leaf blades small, to 25 cm long, densely scaly on underside; plants of granitic rock cliffs and talus; ME, NH, NY, VT . . . . *Dryopteris fragrans*, FRAGRANT WOOD FERN, page 118

1 Leaf blades larger, 25–120 cm long, underside scales sparse or absent ; plants of various habitats but not normally on steep cliffs 2

2 Blades 2-pinnate with the pinnules deeply cut (more than halfway to the base), or 3-pinnate 3

2 Blades 1-pinnate (with the pinnae deeply lobed), to 2-pinnate with the pinnules shallowly cut (halfway or less to the midrib) 5

3 Innermost pinnules (next to rachis) on lower side of lowest pinnae clearly shorter than adjacent pinnule; midribs of pinnae, indusia, and rachis with tiny glands; leaf blades fully evergreen; common, regionwide . . . . *Dryopteris intermedia* . . . . EVERGREEN WOOD FERN, page 122

3 Innermost pinnules (next to rachis) on lower side of lowest pinnae slightly to clearly longer than adjacent lower pinnule; midribs of pinnae, indusia, and rachis without glands; leaf blades not evergreen, dying back in winter 4

4 Leaf blade about as long as the stipe; indusium occasionally glandular; first downward-pointing pinnule of the lowermost pinna 2.5–5 times as long as the first upward-pointing pinnule of the basal pinna; regionwide (absent from RI), especially northward . . . . *Dryopteris campyloptera*, MOUNTAIN WOOD FERN, page 106

4 Leaf blade 2 times as long as the petiole; indusium glabrous; first downward-pointing pinnule of the basal pinna about 2 times as long as the first upward-pointing pinnule of the basal pinna; very common, regionwide . . . . *Dryopteris carthusiana* . . . . SPINULOSE WOOD FERN, page 108

5 Sori very close to margins of blade; base of stipe with many pale brown scales; very common, regionwide . . . . *Dryopteris marginalis*, MARGINAL WOOD FERN, page 124

5 Sori well away from margin of blade, ± in the middle of the pinnules or lobes of the pinnae (or even closer to midvein); stipes with scales either few or dark brown or both 6

6 Stipes very short, 1/5–2/5 length of leaf blade; rachis and costa (midvein of pinna) scaly; plants of dry, rocky forests; uncommon in NH, ME, PA, VT . . . . *Dryopteris filix-mas* . . . . MALE FERN, page 116

6 Stipes 2/5–3/5 length of blade; scales mostly only on stipes; plants of moist forests and swamps 7

7 Lower pinnae triangular, widest near the rachis. 8

7 Lower pinnae ovate, the widest point well away from the rachis, pinnae narrowing both to tip and to base 9

8 Larger leaf blades to 15 cm wide; pinnae of fertile leaf blades tilted horizontally (like the slats of a venetian blind); common, regionwide. . . . . *Dryopteris cristata* . . . . CRESTED WOOD FERN, page 114

8 Larger leaf blades 11–22 (–25) cm wide; pinnae of fertile leaf blades not tilted, the frond essentially flat; regionwide .... *Dryopteris clintoniana*, CLINTON'S WOOD FERN, page 112

9 Leaf blades abruptly narrowed to tip; sori near costa; regionwide .. *Dryopteris goldieana* .... GOLDIE'S WOOD FERN, page 120

9 Leaf blades gradually tapered to tip; sori midway between costa and margin; uncommon in NJ, NY, PA .... *Dryopteris celsa*, LOG FERN, page 110

Dryopteris filix-mas MALE FERN

Dryopteris intermedia EVERGREEN WOOD FERN

Dryopteris campyloptera Clarkson
MOUNTAIN WOOD FERN

DESCRIPTION CT | MA | ME | NH | NJ | NY | PA | VT

FIELD TIPS
- lacy, arching, clumped fern
- sterile and fertile fronds alike, deciduous
- leaf blade broadly triangular
- lower innermost pinnule longer and wider than other pinna; also offset from rachis
- cool, moist woods

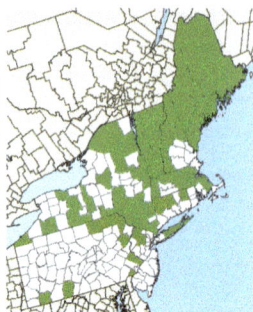

HABITAT

Cool humus-rich acidic soil of wooded rocky slopes, ravines, and swamp margins; southward in its range usually confined to higher elevations.

DESCRIPTION

ROOTSTOCK Dark brown, creeping to semi-erect, covered with brown scales.

FROND Early deciduous, clustered; sterile and fertile fronds alike; to 1m long.

STIPE ca. 2/3 as long as blade, bearing broad brown scales, in part medially dark-striped; grooved, especially upwards; vascular bundles 5.

RACHIS With scattered scales.

BLADE Broadly triangular in outline, light green, 30–60 cm long and 20–40 cm wide, 2-pinnate into 15–20 pairs of pinnae; the pinnules deeply cut into sharply serrate segments.

PINNAE Alternate or opposite; lowest pinna-pair obliquely triangular, the elongate innermost lower pinnule longer than the others and so broad as to match the width of 2 short upper ones, also offset from the rachis (i.e, not directly opposite the innermost upper pinnule); veins free, forked, mostly not extending to margins; margins toothed, tipped with bristles.

SORI Small, in a row between midrib and margin, but sparse throughout the blade; indusium kidney-shaped, brownish, ca. 1 mm wide, glabrous or rarely glandular.

SYNONYMS *Dryopteris austriaca* (Jacq.) Woyn. ex Schinz & Thell., *Dryopteris spinulosa* var. *americana* (Fischer ex Kunze) Fernald

CONSERVATION STATUS *endangered* Pennsylvania.

NOTE Believed to have originated from a cross between *D. expansa* and *D. intermedia.*

NAME Earliest recorded name was *Aspidium spinulosum americanum,* published along with the substitute name of *Aspidium campyloptera* by Kunze in 1848; transferred to genus *Dryopteris* by Clarkson in 1930. From Latin *campylo,* curved, *ptera,* from the Greek *pteron* (wing).

DAVID MCCORQUODALE

JOHN BAUR

Dryopteris campyloptera MOUNTAIN WOOD FERN

Dryopteris carthusiana (Vill.) H.P. Fuchs
SPINULOSE WOOD FERN

DESCRIPTION CT|MA|ME|NH|NJ|NY|PA|RI|VT
FIELD TIPS
- deciduous clumped fern of moist to wet woods
- blade 2-pinnate-pinnatifid, hairs usually absent
- innermost pinnule of basal pinnae longer than next lower pinnules
- indusia kidney-shaped

HABITAT
Moist to wet, deciduous or conifer woods, streambanks, swamp hummocks.

DESCRIPTION
ROOTSTOCK Short-creeping to erect, densely scaly, covered with old stipe bases, forming an irregular clump.
FROND Deciduous, clustered; sterile and fertile fronds nearly alike; sterile fronds tending to be somewhat shorter, arching; to 60 cm long by 15 cm wide.
STIPE Shorter than blade, straw-colored, grooved, base enlarged, light brown scales at base, fewer above; vascular bundles mostly 3–7.
RACHIS With scattered scales.
BLADE Ovate lance-shaped, 2-pinnate-pinnatifid; upper side smooth, lower side smooth or with a few scales or glands.
PINNAE 12 to 14 pairs, held almost horizontally, tips pointing upward; the lowest one not reduced or only slightly reduced in length compared to second pair; first lower pinnule of basal pinna twice as long as upper one, and also larger than second lower pinnule; costae grooved above, continuous from rachis to costae; margins margins serrate, teeth spine-tipped; veins free, forked.
SORI Round, in 1 row between midrib and margin, at the tips of veins, usually absent on the lowest pinna; indusium reniform, gray-white, at a sinus; sporangia dark brown.
SYNONYMS *Dryopteris austriaca* var. *spinulosa* (O.F. Müll.) Fisch., *Dryopteris spinulosa* (O.F. Müll.) Watt

SIMILAR SPECIES
- **Spreading wood fern** (*Dryopteris expansa*) is similar to *D. carthusiana,* but lower basal pinnules of lowest pinnae of D. expansa very large relative to upper basal pinnules.
- Separated from **evergreen wood fern** (*D. intermedia*) by deciduous rather than ever-

green habit, blade generally less dissected, and lower pinnule on the lowest pinna larger than the next one (and that pinnule also attached at or very near point of attachment of first upward pointing pinnule); in *D. intermedia*, this lower inner pinnule on the lowest pinna is somewhat shorter than the next one.

NOTE An early and apparently effective treatment for intestinal tapeworms.

NAME First named in Europe *Polypodium spinulosum* by O. F. Mueller in 1767, and placed in the genus *Dryopteris* by Kuntze in 1891. *Carthusiana,* named after botanist Johan Friedrich Cartheuser (1704–1777). An alternative origin is the village of Carthusium in Dauphiné, France, where Dominique Villars collected it.

WASP32

5173

Dryopteris carthusiana **SPINULOSE WOOD FERN**

Dryopteris celsa (Wm. Palmer) Knowlt., Palmer & Pollard ex Small
LOG FERN

DESCRIPTION NY|PA

FIELD TIPS
- rare deciduous fern of wet places
- similar to *Dryopteris goldieana*, but blade narrower, tapering gradually to tip, and uniformly dark green

HABITAT

Seepage slopes, swamp hummocks and downed logs.

DESCRIPTION

ROOTSTOCK Short-creeping.

FROND Deciduous, clustered; sterile and fertile fronds alike; 120 cm long by 30 cm wide.

STIPE Half or more as long as blade, grooved, bearing mixed broad and narrow pale brown scales with darker central stripe; vascular bundles 3–7 in a C-shaped pattern.

BLADE Ovate lance-shaped, gradually tapered to tip, 1-pinnate-pinnatifid; upper side smooth, lower side with linear to ovate scales.

PINNAE 15 to 20 pairs; basal pinnae linear-oblong, much reduced; costae grooved above, continuous from rachis to costae; margins crenately toothed; veins free, forked.

SORI Round, in 1 row between midrib and margin; indusium kidney-shaped, at a sinus; sporangia brownish.

SYNONYMS *Dryopteris goldieana* subsp. *celsa* Wm. Palmer, *Dryopteris wherryi* F.W. Crane

CONSERVATION STATUS *endangered* New York, Pennsylvania.

SIMILAR SPECIES Clinton's wood fern (*Dryopteris clintoniana*) similar but lowest pinnae wider at base than at middle; in *D. celsa,* lower pinnae narrower at base than middle. **Goldie's wood fern** (*Dryopteris goldieana*) similar but blade in *D. goldieana* tapers abruptly to a short pointed tip, and has dark and light green coloration; blade of *D. celsa* tapers gradually to tip and is uniformly dark green.

NOTE *Dryopteris celsa* is a fertile hybrid between *D. goldieana* and *D. ludoviciana* and is known to hybridize with other species; hybrids can usually be identified by the dark-striped scales. Formerly treated as a subspecies of *Dryopteris goldieana.*

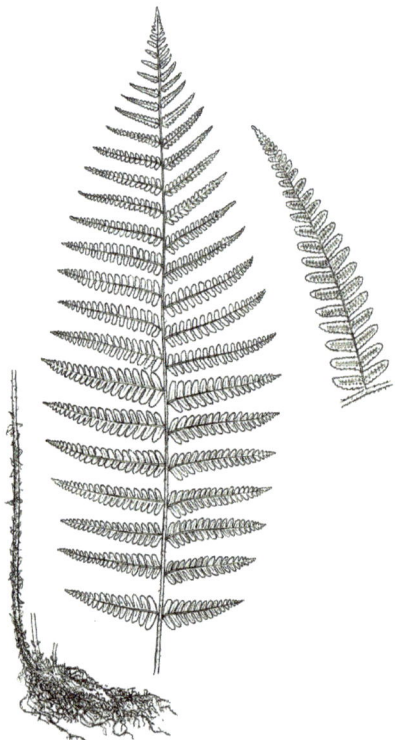

NAME *Celsa* means held high, in reference to the plant's growing perched up on logs or hummocks above water level in swamps.

KRZYSZTOF ZIARNEK

PUMPKIN SKY

Dryopteris celsa **LOG FERN**

Dryopteris clintoniana (D.C. Eat.) Dowell
CLINTON'S WOOD FERN

DESCRIPTION CT | MA | ME | NH | NJ | NY | PA | RI | VT
FIELD TIPS
- loosely clumped fern of wet places
- sterile fronds evergreen, fertile fronds deciduous
- blade with nearly parallel sides, tapering only near tip
- fertile pinnae twisted toward horizontal

HABITAT
Moist to wet woodlands and swamps.

DESCRIPTION

ROOTSTOCK Short-creeping to erect, dark brown to black, with old stipe bases, densely scaly.

FROND Usually evergreen (sterile fronds), borne in 1–2 rows along the rhizome; sterile and fertile fronds nearly alike but fertile fronds deciduous, slightly smaller; 35–80 (–120) cm long by 20 cm wide.

STIPE Shorter than blade, grooved, straw-colored, scaly (at least near base) at base, scales scattered, tan; vascular bundles mostly 3–7 in a C-shaped pattern.

RACHIS Green to pale green, grooved; scaly, especially at base of pinnae.

BLADE In outline, lance-shaped with nearly parallel sides, tapering only in upper 1/4 of blade; 1-pinnate-pinnatifid (nearly more); upper side smooth, lower side with scales.

PINNAE 14 to 16 pairs, more alternate than opposite above the basal pair, less twisted out of plane of blade than in *Dryopteris cristata;* costae grooved above, continuous from rachis to costae; margins sharp-toothed, minutely bristle-tipped; veins free, forked, mostly not reaching margin.

SORI Round, on veins, closer to costa than margin; indusium kidney-shaped, at a sinus; sporangia brownish.

SYNONYMS *Dryopteris cristata* var. *clintoniana* (D.C. Eaton) Underw., *Dryopteris* × *poyseri* Wherry

CONSERVATION STATUS *threatened* Pennsylvania.

SIMILAR SPECIES Similar to crested wood fern (*Dryopteris cristata*) and previously treated as a variety of that species; fronds of *D. clintoniana* tend to be larger, the pinna longer, more narrowly triangular, and less twisted; blade of sterile fronds more abruptly narrowed at tip, similar to Goldie's wood fern.

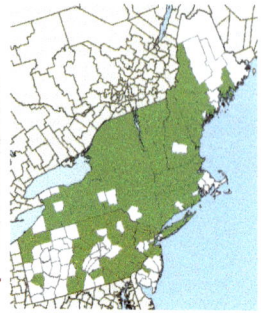

NOTE Originated as a hybrid between *Dryopteris cristata* and *D. goldieana*.

NAME Named by D. C. Eaton *Aspidium cristatum* var. *clintonianum* in 1867. Transferred to *Dryopteris*, though still as a variety, by Underwood in 1893; raised to species rank by Dowell in 1906. Named for Judge G.W. Clinton (1807–1885), a naturalist of Buffalo, New York.

BRAD VON BLON

Dryopteris clintoniana CLINTON'S WOOD FERN

Dryopteris cristata (L.) Gray
CRESTED WOOD FERN

DESCRIPTION CT | MA | ME | NH | NJ | NY | PA | RI | VT
FIELD TIPS
- medium fern of wet places
- fronds narrow
- pinnae short, widely spaced, twisted horizontally
- lowermost pinnae broadly triangular

HABITAT
Hummocks in bogs and sedge meadows, open to semi-shaded places in swamps, wet woods, shrubby wetlands; soils typically acidic.

DESCRIPTION
ROOTSTOCK Short-creeping to erect, stout, dark brown to black, covered with old stipe bases, very scaly.

FROND borne in 1–2 rows along the rhizome; sterile fronds evergreen, somewhat waxy and leathery; fertile fronds less so; sterile fronds shorter and narrower at base than the taller (to 80 cm long), erect fertile fronds.

STIPE Usually shorter than blade, grooved, straw-colored, scaly (at least near base); vascular bundles mostly 3–7 in a C-shaped pattern.

RACHIS Green, with a few scales.

BLADE Narrowly lance-shaped or with parallel sides, 1-pinnate-pinnatifid, often somewhat leathery; upper side smooth, lower side with scales.

PINNAE 10 to 15 pairs, alternate or opposite; fertile pinnae twisted out of plane of blade and perpendicular to it; lower pinnae widely separated, broadly triangular; upwards closer together and narrower; costae grooved above, continuous from rachis to costae; margins spine-toothed; veins free, forked, mostly not reaching margin.

SORI Round, in 1 row between midrib and margin; indusium kidney-shaped, shriveling upon ripening, attached at a sinus; sporangia dark brown.

SIMILAR SPECIES *Dryopteris* × *boottii* (Tuck.) Underw. (*D. cristata* × *D. intermedia*), usually found in swamp forests, in appearance a narrow (and less cut) *D. carthusiana*, but noticeably glandular.

NAME This species was named from European specimens *Polypodium cristatum* by Linnaeus in 1753, and made a *Dryopteris* by Gray in 1848. American plants identical

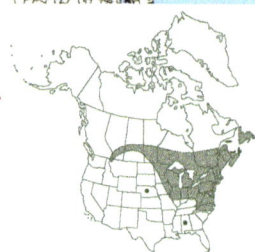

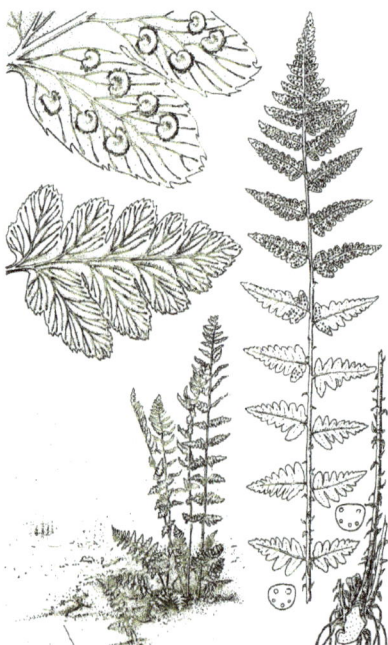

with the European. Latin, *cristata*, like a comb; fertile fronds have the pinnae twisted 90°, and thus appear comblike in profile.

MARINA SADYKOVA

Dryopteris cristata CRESTED WOOD FERN

Dryopteris filix-mas (L.) Schott
MALE FERN

DESCRIPTION ME | NH | PA | RI | VT
FIELD TIPS
- large, dark green, clumped fern of calcareous habitats
- fronds deciduous, 1-pinnate-pinnatifid
- stipe with both wide and hairlike scales

HABITAT
Moist, rocky woods, talus slopes and outcrops; typically over limestone and where shaded, rarely where open.

DESCRIPTION

ROOTSTOCK Erect, thick, dark brown to black, densely scaly, covered with old stipe bases.

FROND Deciduous (sometimes nearly evergreen), clustered, sterile and fertile fronds alike, erect to somewhat arching; 30–120 cm long and 10–25 cm wide.

STIPE Grooved, straw-brown, scales pale brown, of two types - one broad and one hairlike; vascular bundles 5 or 7 in a C-shaped pattern.

RACHIS Green, not grooved or slightly grooved at leaf tip; underside with scales.

BLADE Ovate lance-shaped, widest at middle, strongly narrowed at base; 1-pinnate-pinnatifid to nearly 2-pinnate; herbaceous to somewhat leathery.

PINNAE 16 to 24 pairs, lance-shaped, straight to slightly curved upwards; basal pinnae ovate lance-shaped, reduced in size; basal lower pinnule and basal upper pinnule equal, fully attached to costa along the base; costae grooved above, continuous from rachis to costae; margins serrate but not bristle-tipped; veins free, forked.

SORI Round, in 1 row between midrib and margin, on the upper half of the frond; indusium kidney-shaped, pale green at first, then whitish gray, then rusty brown, then shriveling, at a sinus; sporangia black or dark brown.

SYNONYMS *Aspidium filix-mas* (L.) Swartz, *Lastrea filix-mas* (L.) C. Presl., *Polypodium filix-mas* L., *Polystichum filix-mas* (L.) Roth, *Thelypteris filix-mas* Nieuwl.

CONSERVATION STATUS *endangered* Maine, New Hampshire; *threatened* Vermont.

SIMILAR SPECIES Fronds of marginal wood fern (*Dryopteris marginalis*) more leathery, the stipe longer, sori nearly marginal.

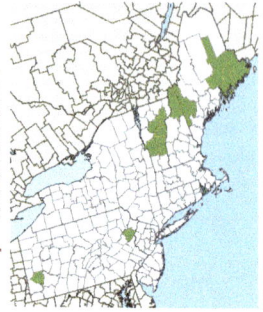

NOTE Previously a popular and effective treatment for tapeworms; the root contains an oleoresin that paralyses tapeworms and other internal parasites; but has since been replaced by less toxic, more effective drugs.

NAME From Latin, *filix,* fern, *mas,* male. Sometimes referred to in ancient texts as 'worm fern'.

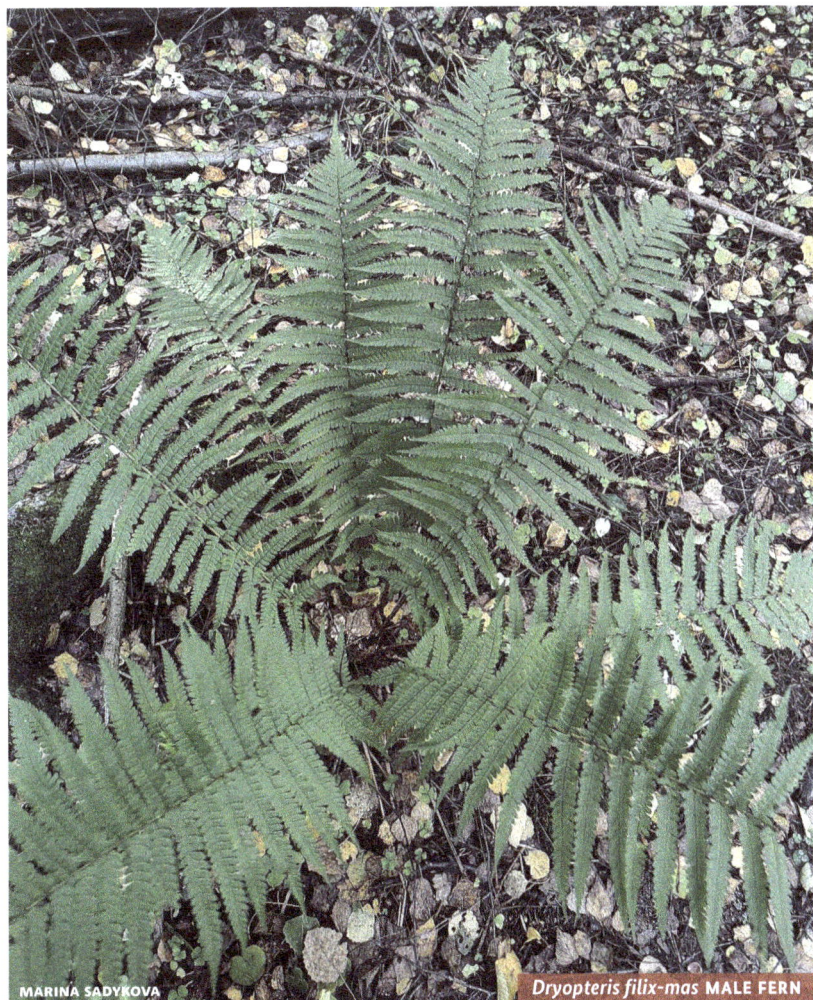

Dryopteris filix-mas MALE FERN

Dryopteris fragrans (L.) Schott
FRAGRANT WOOD FERN

DESCRIPTION ME | NH | NY | VT
FIELD TIPS
- small evergreen fern of dry rock cliffs and boulders
- blade glandular, scented, tapered gradually toward both ends
- curled, dead fronds at base of plant persist for several years

HABITAT
Dry, sunny or partially shaded cliffs, boulders, and talus slopes, on either calcareous or non-calcareous rock.

DESCRIPTION

ROOTSTOCK Erect, short, with old stipe bases and old persistent fronds; covered with brown scales.

FROND Evergreen, clustered; persistent old gray or brown fronds surround base of plant; sterile and fertile fronds alike; usually less than 25 cm long and 5 cm wide.

STIPE Much shorter than blade, grooved, straw-colored, densely scaly; vascular bundles mostly 3–7 in a C-shaped pattern.

RACHIS Glandular, with many reddish-brown scales.

BLADE Narrowly elliptic, gradually tapering toward both ends; pinnate-pinnatifid to 2-pinnate; leathery, with yellow, fragrant glandular hairs; lower side with many brown to reddish scales.

PINNAE Mostly 20 to 30 pairs, opposite, sometimes overlapping, oblong, sessile or short-stalked, lowest pair much shorter than middle pinnae; margins crenately toothed but not bristle-tipped; costae grooved above, continuous from rachis to costae; veins obscure, free, mostly not reaching margin.

SORI Round, large, halfway between midvein and margin; indusium kidney-shaped, glandular, whitish, becoming brown, often overlapping; sporangia chocolate brown.

SYNONYMS *Aspidium fragrans* Sw., *Filix fragrans* Farwell, *Lastrea fragrans* Presl, *Nephrodium fragrans* Richards, *Polypodium fragrans* L., *Polystichum fragrans* (L.) Roth, *Thelypteris fragrans* Nieuwl., *Woodsia xanthosporangia* Ching

CONSERVATION STATUS *threatened* New Hampshire, New York.

SIMILAR SPECIES May be mistaken for **rusty cliff fern** (*Woodsia ilvensis*), but the per-

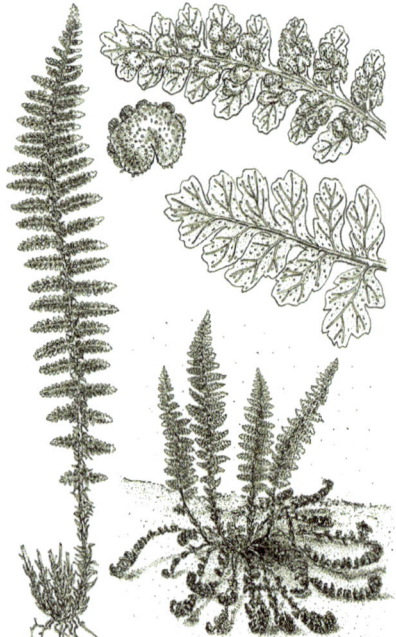

sistent, curled, dead fronds are characteristic of *Dryopteris fragrans*.

NAME Latin, *fragrans,* fragrant; fronds have a sweet, fruity or hay-like odor when crushed.

Dryopteris fragrans **FRAGRANT WOOD FERN**

Dryopteris goldieana (Hook. ex Goldie) Gray
GOLDIE'S WOOD FERN

DESCRIPTION CT | MA | ME | NH | NJ | NY | PA | RI | VT

FIELD TIPS
- region's largest wood fern
- fronds clustered, arched
- blades abruptly tapered at tip
- stipe with glossy scales, dark brown with paler edges
- young fronds usually with faint yellow-green highlights on outer edges

HABITAT

Rich, moist, shaded woods, especially in ravines and near seeps and springs; swamp margins, rocky slopes.

DESCRIPTION

ROOTSTOCK Short-creeping to erect, stout, with old stipe bases, densely scaly.

FROND Deciduous, clustered; sterile and fertile fronds alike, to 120 cm long.

STIPE Shorter than blade, grooved, straw-colored, scaly; scales scattered, dark, glossy brown to nearly black, with pale border; vascular bundles mostly 3–7 in a C-shaped pattern.

RACHIS Green with tan scales; lower portion not grooved, upper segment slightly grooved, with purplish trough.

BLADE Ovate or broadly lance-shaped, tapered abruptly at tip; 1-pinnate-pinnatifid to 2-pinnate-pinnatifid; not glandular, underside scaly, scales absent above.

PINNAE 15 to 20 pairs, lance-shaped, on short stalks, lower pinnae mostly opposite, alternate upwards; lowest pair usually slightly shorter than middle pinnae; costae grooved above, continuous from rachis to costae; margins crenate or serrate, teeth tipped with small bristle; veins free, forked, most not reaching the margin.

SORI Round, in one row near the midvein; indusium kidney-shaped, white to transparent when immature, shriveling, attached at a sinus; sporangia lead gray, then dark brown or black.

SYNONYMS *Aspidium goldieanum* Hooker ex Goldie, *Thelypteris goldieana* (Hooker) Nieuwland, *Nephrodium goldieanum* (Hooker) Hooker & Greville, *Lastrea goldieana* Presl, *Polystichum goldieanum* Keys.

R. A. NONENMACHER

SIMILAR SPECIES Blade of **small log fern** (*Dryopteris celsa*) tapers gradually to tip and is uniformly dark green color; blade in *D. goldieana* tapers abruptly to a short pointed tip, and outer edges often somewhat yellow-green.

NAME First collected by John Goldie, a Scottish botanist (1793–1886), at Montreal, Canada; this fern was named *Aspidium goldianum* by Hooker in 1822, and transferred to *Dryopteris* by Gray in 1848.

EVAN RASKIN

AARONGUNNAR

Dryopteris goldieana **GOLDIE'S WOOD FERN**

Dryopteris intermedia (Muhl. ex Willd.) Gray
EVERGREEN WOOD FERN

DESCRIPTION CT | MA | ME | NH | NJ | NY | PA | RI | VT
FIELD TIPS
- clumped, somewhat arching, evergreen fern
- first downward pointing pinnule on lowest pinna shorter than pinnule next to it
- our only *Dryopteris* with hairs on stipe and rachis (as well as scales)

HABITAT

Moist to dry, deciduous, mixed conifer-hardwood, or coniferous forests, often on slopes and in ravines.

DESCRIPTION

ROOTSTOCK Nearly erect, with old stipe bases, densely scaly.
FROND Evergreen, clustered; sterile and fertile fronds alike; to 70 cm long.
STIPE Shorter than blade, grooved, green to straw-colored, with hairs and tan scales; vascular bundles mostly 3–7 in an arc.
RACHIS With whitish glandular hairs and a few scales.
BLADE Ovate in outline; 3-pinnate at base, gradually less above; glandular.
PINNAE 12 to 20 pairs; lance-shaped and more-or-less in plane of blade, opposite or nearly so, lowest pair not reduced in size; inner lower pinnule on the lowest pinna somewhat shorter than the next one; costae grooved above, continuous from rachis to costae; margins toothed, bristle-tipped; veins free, forked, mostly not reaching margin.
SORI Round, in 1 row between midrib and margin; indusium kidney-shaped, at a sinus, with small glandular hairs; sporangia brown.
SYNONYMS *Dryopteris austriaca* var. *intermedia* (Muhl. ex Willd.) Morton, *Dryopteris spinulosa* var. *intermedia* (Muhl. ex Willd.) Underw.
SIMILAR SPECIES **Spinulose wood fern** (*Dryopteris carthusiana*) often confused with *D. intermedia,* but in most plants, lower basal pinnule of the lowest pinna is longer than the adjacent pinnules; in *D. intermedia* lower basal pinnule is shorter than the adjacent pinnules. Fronds of *D. intermedia* evergreen and those of *D. carthusiana* are not, helpful especially in winter and spring. Glands on indusia and

midveins of pinnules of *D. intermedia* are present, but can be difficult to see, requiring a hand lens, or are shed with age; no glands on *D. carthusiana.*

NOTE *Dryopteris intermedia* known to hybridize with 8 species; hybrids all have distinctive glandular hairs on the indusia and usually also on costae.

NAME Discovered by Muhlenberg in Lancaster Co., Pennsylvania. His manuscript name for it, *Aspidium intermedium,* was published by Willdenow in 1810, and transferred to *Dryopteris* by Gray in 1848. From Latin *intermedius,* presumably intermediate between *D. carthusiana* and *D. campyloptera.*

DOUG MCGRADY

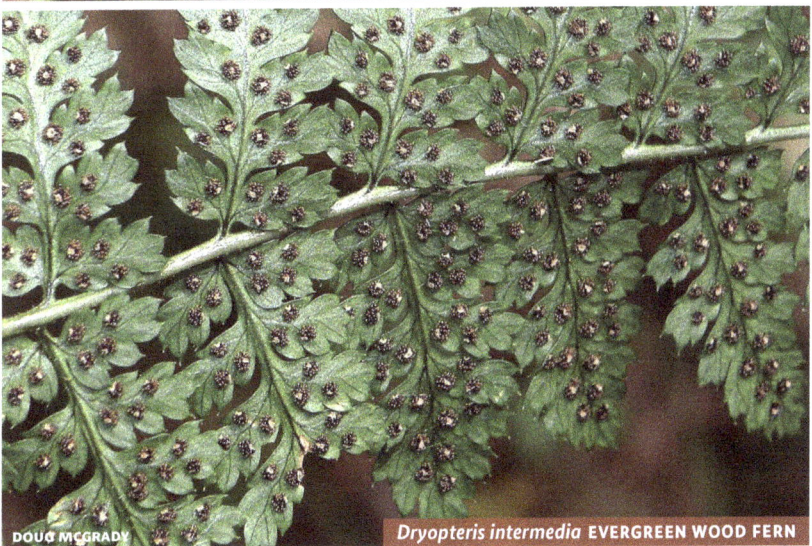
DOUG MCGRADY

Dryopteris intermedia **EVERGREEN WOOD FERN**

Dryopteris marginalis (L.) Gray
MARGINAL WOOD FERN

DISTRIBUTION CT | MA | ME | NH | NJ | NY | PA | RI | VT

FIELD TIPS
- large evergreen fern, usually where rocky
- fronds leathery, sometimes blue-green; dead fronds persist at base of plant
- pinule margins smooth or round- toothed
- stipes very scaly
- sori very near margins, in upper portion of blade

HABITAT

Rocky wooded slopes and ravines, streambanks, usually in partial shade; soils alkaline to acidic.

DESCRIPTION

ROOTSTOCK Upright, thick, with old stipe bases; densely covered with gold-brown chaffy scales.

CROZIER Densely covered with golden brown hairs.

FROND Evergreen, clustered; sterile and fertile fronds alike; to 100 cm long or more.

STIPE Shorter than blade, grooved, reddish brown at swollen base, straw-colored above, densely covered with tan scales; vascular bundles mostly 3–7 in a C-shaped arc, sometimes fewer at stipe apex.

RACHIS Pale to light green, narrowly grooved; scaly, especially toward base of pinnae.

BLADE Ovate lance-shaped; 2-pinnate; leathery, deep green to felty blue-green; underside scaly, upper side without scales.

PINNAE 12 to 16 pairs, lance-shaped, short-stalked, lowest pair usually a little shorter than middle pinnae; basal pinnules longer than adjacent pinnules, lower pinnule longer than adjacent upper pinnule; costae grooved above, continuous from rachis to costae; margins shallowly round-toothed to nearly entire; veins free, forked, mostly not reaching margin.

SORI Round, in 1 row near the margin on the top two-thirds of the blade; indusium kidney-shaped, silvery, attached at a sinus; sporangia lead gray, maturing to dark brown.

SIMILAR SPECIES In **male fern** (*Dryopteris filix-mas*), sori not marginal, usually more than 20 pairs of pinnae, with lowest pair much shorter than middle pinnae.

NAME Sent to Linnaeus from Canada and named by him *Polypodium marginale* in

1753; assigned to *Dryopteris* by Gray in 1848. From Latin, *marginatus*, enclosed with a margin, referring to position of sori.

Dryopteris marginalis MARGINAL WOOD FERN

Polystichum Roth HOLLY FERN, SWORD FERN

LEATHERY, EVERGREEN FERNS that are generally monomorphic (except in *P. acrostichoides* with smaller fertile pinnae). Habitats vary from forests to rock cliffs and talus. Worldwide, *Polystichum* includes about 500 species, 2 species in the region.

KEY CHARACTERS

- Similar to *Dryopteris*, but distinguished by the upward ear on the pinnae, commonly spiny margins, the peltate (round) indusia, and discontinuous grooving between rachis and costa. Also similar to the introduced *Cyrtomium*, but that genus has netted veins (and in the northeast only reported from MA and NJ); veins in *Polystichum* are free and not net-like.
- Fronds evergreen, sterile and fertile fronds alike except in *P. acrostichoides;* stipes scaly; leaf blades 1–3 pinnate, lance-shaped, glossy; pinnae usually with an upper side lobe (auricle), costae grooved above, grooves discontinuous from rachis to costae.
- Sori round, in 1 (2) rows between midrib and margin, indusium peltate or rarely absent.

NAME From Greek, *polys,* many, and *stichos,* row, referring to the several parallel rows of sori.

KEY TO *POLYSTICHUM* | HOLLY FERN, SWORD FERN

1 Pinnae of two types, terminal fertile pinnae distinctly narrower and smaller than sterile pinnae; leaf blade evergreen; common, regionwide . . . . *Polystichum acrostichoides*
. . . . CHRISTMAS FERN, page 128
1 Pinnae uniform; fertile and sterile pinnae similar in size and shape; CT, MA, ME, NH, NY, PA, VT. . . . . *Polystichum braunii*, BRAUN'S HOLLY FERN, page 130

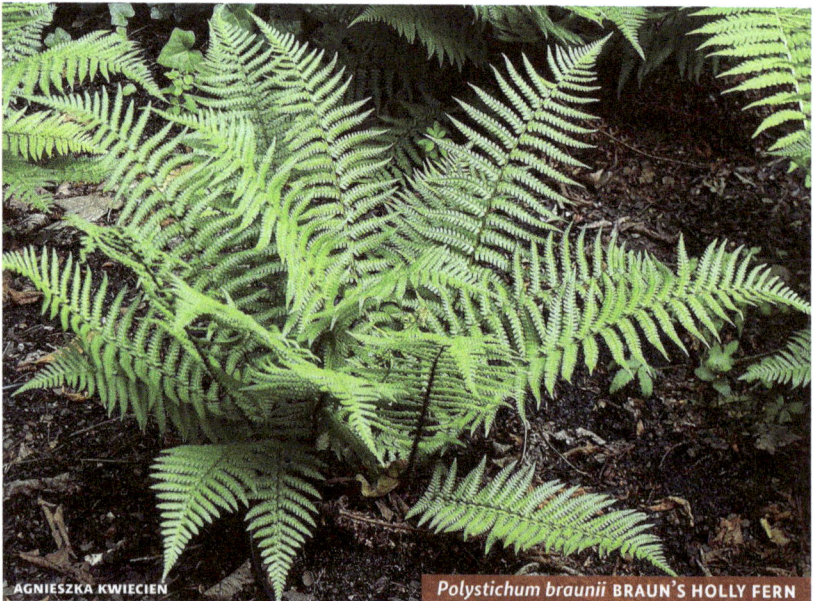

AGNIESZKA KWIECIEN
Polystichum braunii BRAUN'S HOLLY FERN

Polystichum acrostichoides CHRISTMAS FERN

MIPAPAY

Polystichum acrostichoides (Michx.) Schott
CHRISTMAS FERN

DISTRIBUTION CT | MA | ME | NH | NJ | NY | PA | RI | VT

FIELD TIPS
- clumped evergreen fern
- fronds dark green, satiny
- fertile pinnae reduced in size, on upper portion of blade
- pinnae with upward-pointing lobe
- stipe scaly

HABITAT
Shaded, moist to dry forests, rocky slopes and ravines; soils acidic to neutral.

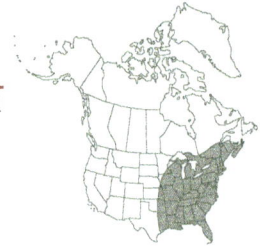

DESCRIPTION

ROOTSTOCK Ascending or creeping, with old stipe bases and wilted fronds attached, pale brown, densely scaly.

CROZIER Bent backwards, 1 cm wide, covered with silvery-white scales.

FROND Evergreen, clustered; sterile and fertile fronds somewhat different, those with fertile pinnae taller and more rigid than the completely sterile fronds, the fertile pinnae reduced in size and only in upper portion of blade; fronds to 80 cm long or more.

STIPE Grooved, green, densely scaly; scales light brown, smaller in size upwards; vascular bundles 5 in an arc.

RACHIS Green, grooved, very scaly.

BLADE Linear lance-shaped, base narrowed, 1-pinnate, glossy; underside with numerous small, fine fine scales.

PINNAE 20 to 30 pairs, lower pinnae bend forward and down; upper auricle well developed; margins entire to toothed and spiny; veins free, forked.

SORI Distinct, green at first, confluent later, completely covering lower surface of fertile pinnae (on the upper, reduced pinnae only); indusium round, attached at center, shriveling with age; sporangia golden brown.

SIMILAR SPECIES Fronds of **glade fern** (*Homalosorus pycnocarpos*) also 1-pinnate, but that species deciduous and pinnae without upward-pointing lobes

NOTE Used by early New England settlers as a Christmas decoration.

NAME Michaux observed this fern during his explorations in "Pennsylvania, Carolina, and Tennessee," and named it *Nephrodium acrostichoides* in 1803. It was

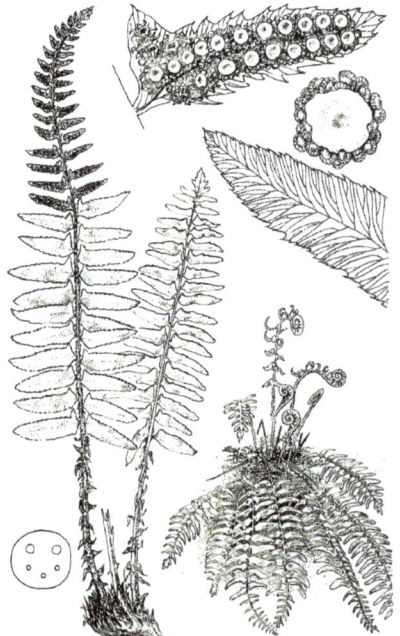

transferred to *Polystichum* by Schott in 1834. *Acrosti-choides* means resembling *Acrostichum,* a tropical genus, where the sori completely cover the pinna underside.

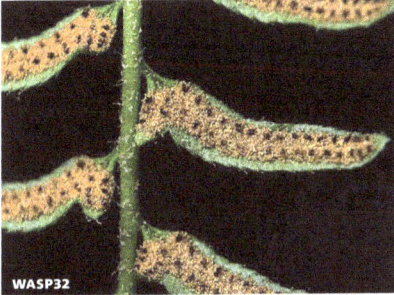

WASP32

LAURA CLARK

SANDY WOLKENBERG

Polystichum acrostichoides **CHRISTMAS FERN**

Polystichum braunii (Spenner) Fée
BRAUN'S HOLLY FERN

DISTRIBUTION CT | MA | ME | NH | NY | PA | VT

FIELD TIPS
- clumped, arched, semi-evergreen fern
- fronds dark green, shiny, tapered to tip and base, lower pinnae very small
- stipe and rachis densely scaly
- pinnule margins with bristle-tipped teeth
- in rocky woods in mostly northern portions of region

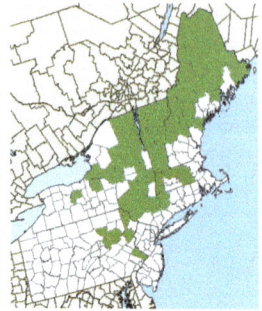

HABITAT

Cool, moist, shaded woods; especially where rocky; moist cliffs and boulders; soils circumneutral.

DESCRIPTION

ROOTSTOCK Erect, stout, with old stipe bases, very scaly; fronds arise in vase-like cluster.

CROZIER For next year's fronds formed in late summer, these covered with silvery to light brown scales.

FROND Evergreen (or nearly so), dark green, clustered; sterile and fertile fronds alike, to 100 cm long.

STIPE Grooved; very scaly, the scales large and also hairlike, silvery at first, then light brown; vascular bundles 5–7 in an arc.

RACHIS Light brown, grooved, densely scaly.

BLADE Ovate, broadest at middle, tapered gradually to base, lowest pinnae only 2 cm long; upper pinnae lobed, lower pinnae divided to costa to form 6–18 pairs of pinnules; dark green, leathery; with whitish to brown hairlike scales on both sides.

PINNAE 20 to 40 pairs, oblong; pinnules short-stalked, eared near the rachis, the first upper pinnule the same size or only slightly larger than the next; margins toothed, bristle-tipped; veins free, forked.

SORI Round, distinct and separate, in 1 row between midrib and margin; indusium round, attached at center, light brown, shriveling with age; sporangia gray-brown to black.

CONSERVATION STATUS _endangered_ Massachusetts, Pennsylvania.

NOTE One of our most attractive ferns.

NAME Discovered in Europe, and named _Aspidium braunii_ by Spenner in 1825, then transferred to genus _Polystichum_ by Fée

around 1850. In honor of Alexander Carl Heinrich Braun, 1805–1877, German professor of botany.

Polystichum braunii **BRAUN'S HOLLY FERN**

EQUISETACEAE *Horsetail family*

Equisetum L. HORSETAIL, SCOURING-RUSH

EVERGREEN OR DECIDUOUS PERENNIALS from creeping rhizomes. Modern classifications describe a single extant genus within the family and 15 species, with nine species in our flora; hybrids also occur.

KEY CHARACTERS

• Stems rush-like, jointed, sometimes hollow; branched or unbranched; internodes of stems commonly ridged longitudinally, with stomata in rows or bands in the grooves, and the ridges with siliceous bumps or bands, giving stems of some species (the scouring rushes) a gritty feel.

• Leaves small, whorled, fused into a collarlike sheath at each node; teeth on the sheath represent tips of the leaves; when present, the branches are slender extensions of the sheaths.

• Spores green, spherical, wrapped with 4 elaters, borne in sporangia on sporophylls in cones. The hexagonal plates on the cone separate when the spores ready to be shed; spores green and viable for only a few days.

• Cones terminal on vegetative stems, or in some species on specialized early season shoots that lack chlorophyll.

NAME The common name **scouring rush** was given to unbranched species of *Equisetum* by American pioneers who used the silica-roughened stems to scrub pots and pans. Before commercial scouring powders were developed for cleaning pots and pans, pulverized stems of these species were sometimes used. The common name **horsetail** was given to species with whorled branches, in reference to their bushy, bristly appearance resembling a horse's tail. The genus name from Greek, *equis,* horse, and *seta,* bristle.

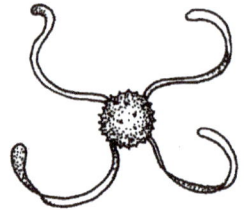

Equisetum arvense
spore elators
(elators help with
spore dispersal)

EQUISETUM HYBRIDS

• *Equisetum* × *ferrissii* Clute: *Equisetum praealtum* and *E. laevigatum.*

• *Equisetum* × *litorale* Kuehl ex Rupr.: *Equisetum arvense* and *E. fluviatile.*

• *Equisetum* × *mackayi* (Newm.) Brichan: *E. praealtum* and *E. variegatum*; CT, ME, NH, northern NY, VT.

• *Equisetum* × *nelsonii* (A.A. Eat) Schaffn.: *E. laevigatum* and *E. variegatum.*

KEY TO *EQUISETUM* | HORSETAIL, SCOURING-RUSH

1 Stems evergreen (annual in *E. laevigatum*); unbranched or with a few scattered branches, branches not in regular whorls (**scouring rushes**) 2

1 Stems deciduous; usually with regular whorls of branches, sometimes unbranched (**horsetails**) 5

2 Stems solid (central cavity absent); stems small, slender and sprawling; CT, MA, ME, NH, NY, northwest PA, VT . . . . *Equisetum scirpoides*, DWARF SCOURING-RUSH, page 146

2 Stems hollow (central cavity present); stems larger, usually upright 3

3   Stems 1–3 dm tall, with 5–12 ridges, central cavity to 1/3 diameter of stem; regionwide .
    . . . . *Equisetum variegatum*, VARIEGATED SCOURING-RUSH, page 150

3   Stems usually taller, with 16–50 ridges, central cavity more than half diameter of stem . **4**

4   Cones tipped with a distinct, small sharp point; stem sheaths with a black band at tip and
    base; regionwide. . . . . *Equisetum praealtum*, COMMON SCOURING-RUSH, page 142

4   Cones blunt-tipped, sheaths with black band at tip only; uncommon in PA . . . .
    . . . . *Equisetum laevigatum*, SMOOTH SCOURING-RUSH, page 138

5   Stems unbranched **6**

5   Stems with regular whorls of branches **10**

6   Stems green **7**

6   Stems brown or flesh-colored (fertile stems) **8**

7   Stems with 9–25 shallow ridges; central cavity more than half diameter of stem; sheath
    teeth entirely black or with narrow white margins; regionwide . . . . . *Equisetum fluviatile*
    . . . . WATER HORSETAIL, page 136

7   Stems with 5–10 strongly angled ridges; central cavity less than 1/3 diameter of stem;
    sheath teeth with white margins and dark centers; CT, MA, ME, NH, NJ, NY, VT . . . .
    . . . . *Equisetum palustre*, MARSH HORSETAIL, page 140

8   Sheath teeth papery and red-brown, teeth joined and forming several broad lobes; com-
    mon, regionwide . . . . *Equisetum sylvaticum*, WOODLAND-HORSETAIL, page 148

8   Sheath teeth black or brown, not papery, separate or joined in more than 4 small groups
    . . . . **9**

9   Fertile stems withering after spores mature, remaining unbranched; common, regionwide
    . . . . *Equisetum arvense*, FIELD HORSETAIL, page 134

9   Fertile stems persistent, becoming branched and green; CT, MA, ME, NH, NJ, NY, VT . . . .
    . . . . *Equisetum pratense*, MEADOW HORSETAIL, page 144

10 First internode of each branch shorter than the subtending sheath of the main stem . . **11**

10 First internode of each branch equal or longer than the subtending sheath of the main
    stem **12**

11 Stems with 9–25 shallow ridges; central cavity more than half diameter of stem; sheath
    teeth more than 12, entirely black or with narrow white margins; regionwide . . . .
    . . . . *Equisetum fluviatile*, WATER HORSETAIL, page 136

11 Stems with 5–10 strongly angled ridges; central cavity about same size as outer cavities;
    sheath teeth 5–6, with white margins and dark centers; CT, MA, ME, NH, NJ, NY, VT . . . .
    . . . . *Equisetum palustre*, MARSH HORSETAIL, page 140

12 Stem branches themselves branched; sheath teeth papery and red-brown, teeth joined to
    form several broad lobes; common, regionwide . . . . *Equisetum sylvaticum*
    . . . . WOODLAND HORSETAIL, page 148

12 Stem branches unbranched; sheath teeth black or brown, not papery, separate or joined
    in more than 4 small groups **13**

13 Stem branches ascending; teeth of branch sheaths gradually tapering to a slender tip; com-
    mon, regionwide . . . . *Equisetum arvense*, FIELD HORSETAIL, page 134

13 Stem branches spreading; teeth of branch sheaths broadly triangular; CT, MA, ME, NH, NJ,
    NY, VT . . . . *Equisetum pratense*, MEADOW HORSETAIL, page 144

Equisetum arvense L.
FIELD HORSETAIL

DISTRIBUTION CT | MA | ME | NH | NJ | NY | PA | RI | VT
FIELD TIPS
• very common, colony-forming
• stems deciduous
• sterile stems with dense whorls of branches from the nodes
• fertile stems smaller, nongreen, and appear in spring before sterile stems, then withering soon after shedding spores
• cones rounded at tip

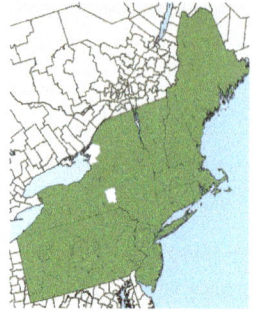

HABITAT
Variety of sunny or shaded habitats (apart from where extremely wet or dry), including fields, woods, shores; often in somewhat disturbed places such as ditches, roadsides, railroad banks; sometimes weedy in yards.

DESCRIPTION
ROOTSTOCK Dark brown to black, hairy, sometimes bearing tubers; colony-forming.
STEMS Deciduous, of two kinds, sterile and fertile, the sterile stems appearing as the fertile stems wither.
STERILE STEMS Upright to prostrate, 10–60 cm tall, 3–5 mm thick; central cavity usually less than 1/2 stem diameter (to sometimes slightly more than 1/2), with a ring of smaller cavities alternating with the 4–14 ridges; silica in dots on the ridges; sheaths with 4–14 short, narrow, dark, chaffy-margined teeth, these separate or occasionally joined; branches solid, in regular whorls, spreading or ascending, the branches themselves mostly unbranched, 3- or 4-angled in cross-section; the first internode of branch longer than the corresponding stem sheath.
FERTILE STEMS Without chlorophyll, appearing in spring before sterile stems (and conspicuous because of their numbers along some roadsides), about 10–30 cm tall and generally shorter than the sterile stems, pale brown or pinkish, fleshy, unbranched, with 4–6 pale brown sheaths, these tipped with 6–12 darker teeth; dying after the spores are shed in spring
CONES Long-peduncled, rounded on top and not tipped with a sharp point.
SYNONYMS *Equisetum calderi* B. Boivin
SIMILAR SPECIES Sterile stems of *Equisetum arvense* are perhaps most frequently confused with *E. pratense* and *E. palustre*:
• **Meadow horsetail** (*Equisetum pratense*) is more delicate in aspect, and the stems are whitish green;

ABOVE stem cross-section
RIGHT fertile stem

also, the teeth are traingular rather than narrowly lance-shaped as in *E. arvense.*

• In *E. arvense* the first internode of the branch is longer than the subtending teeth, whereas in **marsh horsetail** (*E. palustre*) the first internode is shorter than the subtending teeth.

NAME Named by Linnaeus in 1753, with reference to occurrences in Europe and Virginia. From Latin, *arvum,* field or cultivated land.

JANE RICHARDSON

Equisetum arvense FIELD HORSETAIL

Equisetum fluviatile L.
WATER HORSETAIL

DISTRIBUTION CT | MA | ME | NH | NJ | NY | PA | RI | VT

FIELD TIPS
- colony-forming, shallow water and muddy shores
- stems deciduous, unbranched or with small branches
- large central cavity, stems easily crushed when squeezed
- cones rounded at tip

HABITAT

In quiet, shallow water to about 60 cm deep of ponds, lakes and rivers, wet shores, ditches, bog margins, typically in full sun.

DESCRIPTION

ROOTSTOCK Smooth, shiny, light brown to reddish, branching; the stems single but often forming dense colonies from the spreading rhizomes.

STEMS Deciduous, 30–150 cm tall, 3–8 mm thick; central cavity more than 4/5 diameter of stem; smaller outer cavities small or absent; 10–30 smooth ridges present; sheaths tightly appressed, with 15–20 teeth; teeth dark brown, narrow, acuminate, persistent. Stems unbranched, or with branches occurring sporadically, or sometimes whorled; branches to 15 cm long, hollow, with 4–6 ridges, the first internode shorter than the stem sheath.

CONES To 2.5 cm long, yellow to brown, rounded at top, stalked, deciduous, shedding spores late spring through the summer.

SYNONYMS *Equisetum limosum* L.

NOTE A hybrid with *Equisetum arvense* (*E. × litorale* Kuehl ex Rupr.) is frequent in places where both species occur.

NAME This plant in Europe, as in America, occurs in both a branchless form and one with conspicuous whorls of branches; not recognizing their identity, Linnaeus in 1753 applied the name *limosum* to the first and *fluviatile* to the second. The latter was validated by Ehrhart in 1792. From Latin, *fluviatile,* pertaining to a river.

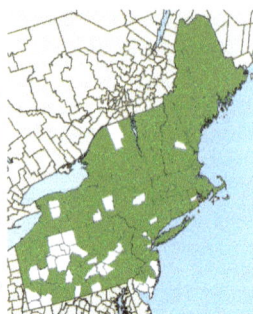

stem cross-section

ALGIRDAS

DARIUSZ KOWALCZYK

STANISLAV MURASHKIN

Equisetum fluviatile WATER HORSETAIL

Equisetum laevigatum A. Braun
SMOOTH SCOURING-RUSH

DISTRIBUTION PA

FIELD TIPS
- common, colony-forming; open, sandy places
- stems smooth, deciduous, unbranched
- sheaths green with a black rim; sheath teeth early deciduous
- cones blunt-tipped

HABITAT

Moist to dry sandy river terraces, shores and meadows, open swamps, moist dunes, prairies, ditches, roadsides, along railways; usually where open or partially shaded.

DESCRIPTION

ROOTSTOCK Thick, dull, rough, dark brown; stems single or several together from the creeping rhizome.

STEMS Deciduous, light-green, 40–80 cm tall, 2–7 mm thick, usually unbranched; central cavity 2/3 to 3/4 diameter of stem; smaller outer cavities alternating with 14–26 ridges; ridges rounded, smooth or with cross-bands of silica; sheaths constricted at the base, flaring towards the top, same color as the stem except for narrow band at top, or in old stems the lower sheath becoming girdled with brown; teeth narrowly lance-shaped, dark with chaffy margins, soon deciduous and leaving only small depressions at the sheath tips.

CONES To 2 cm in length when mature, yellow to brown, short-stalked, the tip with a small sharp point or rounded; shedding spores from late spring into summer.

SYNONYMS *Equisetum funstonii* A.A. Eaton, *Equisetum kansanum* Schaffn., *Hippochaete laevigata* (A. Braun) Farw.

SIMILAR SPECIES **Tall scouring-rush** (*Equisetum praealtum*) is similar, but has two dark bands on each leaf sheath, and often found in somewhat wetter habitats.

NAME Named in 1844 by A. Braun from specimens collected along banks of Mississippi River south of St. Louis, Missouri. From Latin, *laevigatum,* smooth, referring to the relatively smooth stems.

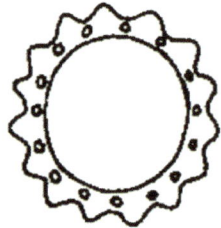

stem cross-section

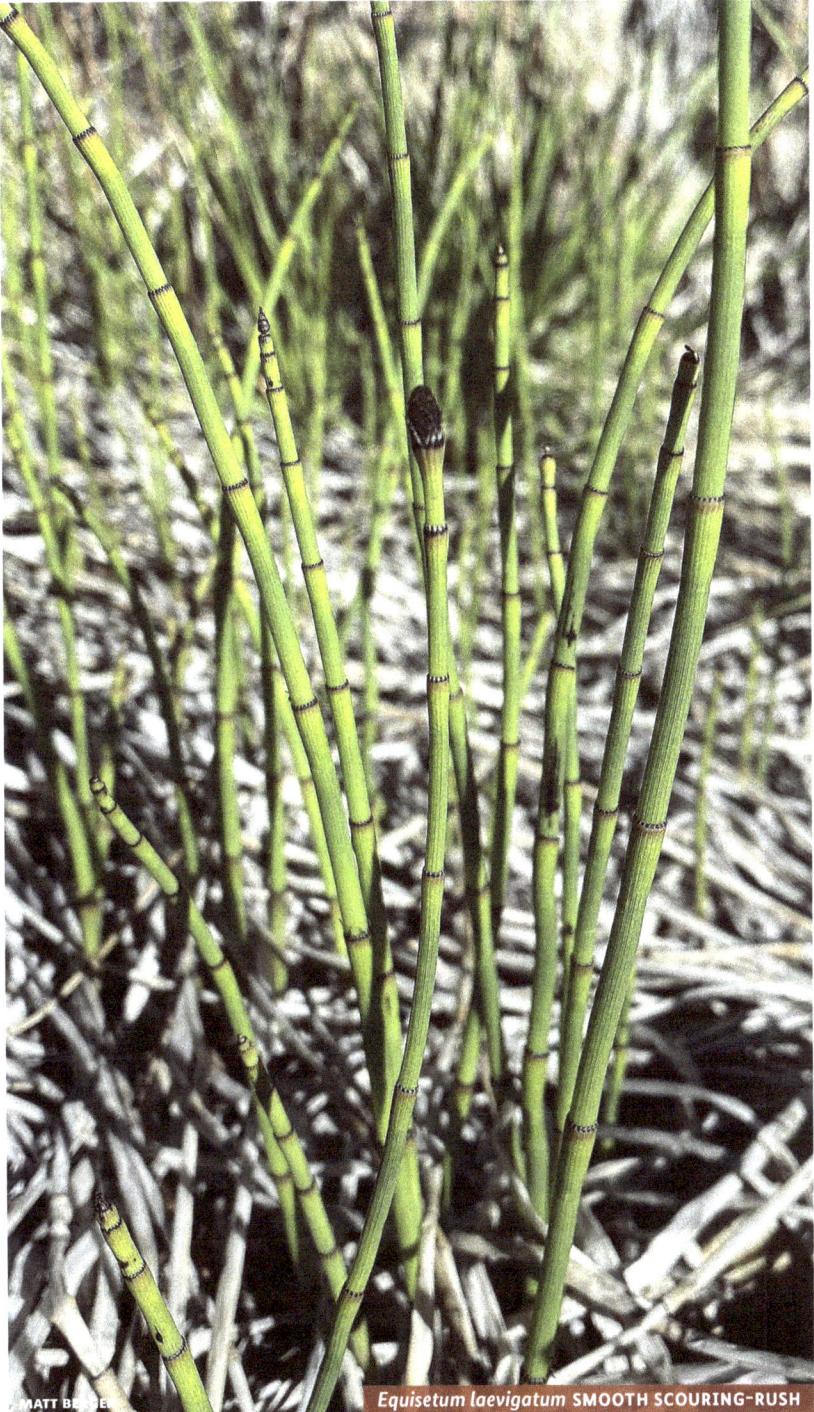

MATT BE...

Equisetum laevigatum SMOOTH SCOURING-RUSH

Equisetum palustre L.
MARSH HORSETAIL

DISTRIBUTION CT | MA | ME | NH | NJ | NY | VT

FIELD TIPS
- in shallow water or wet places
- stems deciduous, branched
- small central cavity
- cones blunt-tipped

HABITAT
Wet woods and meadows, swamps, ditches, shores, cold water seeps, soils wet or inundated up to about 20 cm deep.

DESCRIPTION

ROOTSTOCK Black to brown, shiny, occasionally tuber-bearing; stems single or clustered from the rhizomes.

STEMS Deciduous, mostly 20–60 cm long, 1–3 mm thick, erect, solitary or clustered; central cavity less than 1/2 the stem diameter; outer cavities about same size as central cavity and alternating with 5–10 prominently angled smooth or rough ridges; sheaths green with long, narrow, black and chaffy-margined teeth. Branches spreading in regular whorls from the middle nodes (sometimes few to none), with the first branch internode shorter than the subtending stem sheath.

CONES 1–3.5 cm long, rounded at tip, deciduous, on a stalk at end of main stems; spores shed in summer.

CONSERVATION STATUS *endangered* New Hampshire; *threatened* New York, Vermont.

NOTE Marsh horsetail usually found in wetter situations than the similar **field horsetail** (*Equisetum arvense*).

NAME In 1753 Linnaeus gave the name *Equisetum palustre* to a European plant; it was later extended to include similar American specimens. From Latin, *palustris,* marshy.

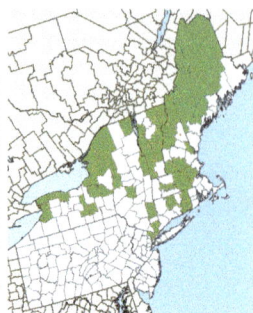

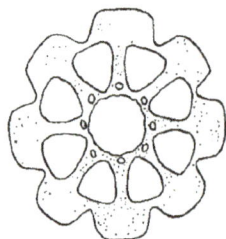

stem cross-section

VLADIMÍR FUKA

KRZYSZTOF ZIARNEK

PATRICK MEURIN

Equisetum palustre MARSH HORSETAIL

Equisetum praealtum Raf.
TALL SCOURING-RUSH

DISTRIBUTION CT | MA | ME | NH | NJ | NY | PA | RI | VT

FIELD TIPS
- common, colony-forming, sandy places
- stems evergreen and rough, unbranched
- sheaths with wide, ash-gray band bordered by a lower black band and an upper black rim
- cones tipped by a small point

HABITAT

Sandy meadows and old fields, shores, dunes, roadsides, ditches, railroad embankments; sites wet to dry and in sun or shade.

DESCRIPTION

ROOTSTOCK Thick, blackish or dark brown, dull, rough, the stems arising singly or several together from the deep rhizome, and sometimes forming large colonies.

STEMS Evergreen, 30–120 cm long, 4–6 mm thick, dark green, usually unbranched; central cavity more than 2/3 diameter of stem; small outer cavities alternating with 14–50 ridges; ridges broad, flat, or rounded, with prominent cross-bands; sheaths constricted at base, same color as the stem when young, but soon developing dark bands at base and top, with area between base and summit white or ashy gray; teeth lance-shaped, dark brown with broad chaffy margins, usually soon deciduous, leaving only darker traces.

CONES To 2 cm long when mature, yellow to black, short-stalked, tipped by a small point, shedding spores from summer to early fall, or persisting unopened until the following spring.

stem cross-section

SYNONYMS *Equisetum affine* Engelm., *Equisetum hyemale* L., *Equisetum robustum* A. Braun, *Hippochaete hyemalis* (L.) Bruhin

NOTE The rough stems of tall scouring rush have been used to scour or clean pots, and as a sandpaper.

NAME North American plants now considered distinct from the **European scouring-rush** (*Equisetum hyemale*). Named *Equisetum praealtum* by Rafinesque in 1817, with reference to colonies along the Mississippi River in Louisiana.

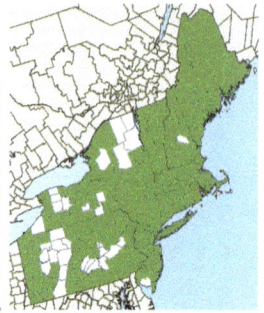

PATRICK HANLY

KRZYSZTOF ZIARNEK

Equisetum praealtum TALL SCOURING-RUSH

Equisetum pratense Ehrh.
MEADOW HORSETAIL

DISTRIBUTION CT | MA | ME | NH | NJ | NY | VT

FIELD TIPS
- colony-forming horsetail of wet places
- stems deciduous, branched, either sterile or fertile, the fertile stems seldom produced
- sheath teeth white-margined, persistent
- cones rounded at tip

HABITAT

Conifer swamps, moist to wet deciduous or mixed conifer-hardwood forests, often near springs or seeps; in sun or partial shade.

DESCRIPTION

ROOTSTOCK Dull, black; the stems mostly growing singly from the rhizome.

STEMS Deciduous, of two kinds, sterile and fertile; the fertile stems rarely seen.

STERILE STEMS Whitish green, to 60 cm tall and 1–3 mm thick, regularly and abundantly branched; central cavity less than 1/2 diameter of stem, with small outer cavities alternating with the 8–18 ridges; silica spicules in 3 rows on the ridges of the middle and upper internodes; sheaths pale, persistent, with 10–20 narrow teeth, the teeth dark with white margins; branches in regular whorls, horizontal to drooping, triangular in cross-section, the first internodes of each branch shorter than the corresponding sheaths; the branch teeth slightly incurved, with thin white margins.

FERTILE STEMS Apparently not common, at first unbranched and lacking chlorophyll, fleshy, becoming green and branched after the spores are shed; sheaths and teeth about twice as long as those of the sterile stems.

CONES Peduncled, to 3 cm long, rounded at the tip, deciduous, shedding spores from late-spring into summer.

CONSERVATION STATUS *endangered* Connecticut, New Jersey; *threatened* New York.

SIMILAR SPECIES Separated from **field horsetail** (*Equisetum arvense*) by the fertile stems being persistent, turning green, and developing branches.

NAME Well-known in northern Europe, this horsetail was named by Ehrhart in 1784. From Latin, *pratum,* meadow, referring to its habitat.

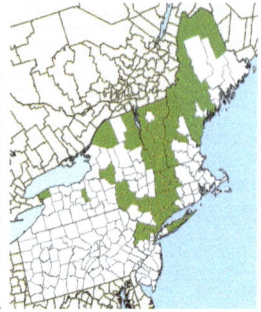

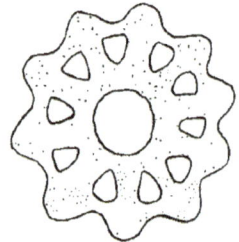

stem cross-section

VALERII GLAZBNOV   *Equisetum pratense* MEADOW HORSETAIL

Equisetum scirpoides Michx.
DWARF SCOURING-RUSH

DISTRIBUTION CT | MA | ME | NH | NJ | PA | VT

FIELD TIPS
- plants small, evergreen, clumped
- stems kinked at the nodes; each stem node with 3 teeth
- stems solid, not hollow, unique among our *Equisetum*
- cones small, tipped with a small point

HABITAT

Cedar swamps, moist mixed forests, usually in shade; sometimes on open north-facing bluffs and ledges; may form mats, the stems partly buried in humus.

DESCRIPTION

ROOTSTOCK Shallow, branching, the stems in tufts and forming colonies via the spreading rhizomes.

STEMS Evergreen, mostly 10–20 cm long, slender, 0.5–1.0 mm thick, dark green, usually unbranched, ascending or prostrate, arched and recurving; center of stem solid, the central cavity absent, replaced by 3 (rarely 4) cavities alternating with the deeply grooved ridges; silica rosettes in lines on the crests of the ridges; sheaths green below, black above, loose, with 3 (rarely 4) triangular teeth; teeth dark with white margins, persistent, but their awl-like tips usually soon breaking off.

CONES Small, 2–3 mm long, black, tipped by a small point, shedding spores in summer, or persisting unopened until the following spring.

CONSERVATION STATUS *endangered* Connecticut, Pennsylvania.

NAME From Latin, *scirpus,* rush or bulrush, Greek *eidos,* like; the small tufted stems similar to a rush.

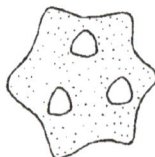

stem cross-section

JON BOUTON

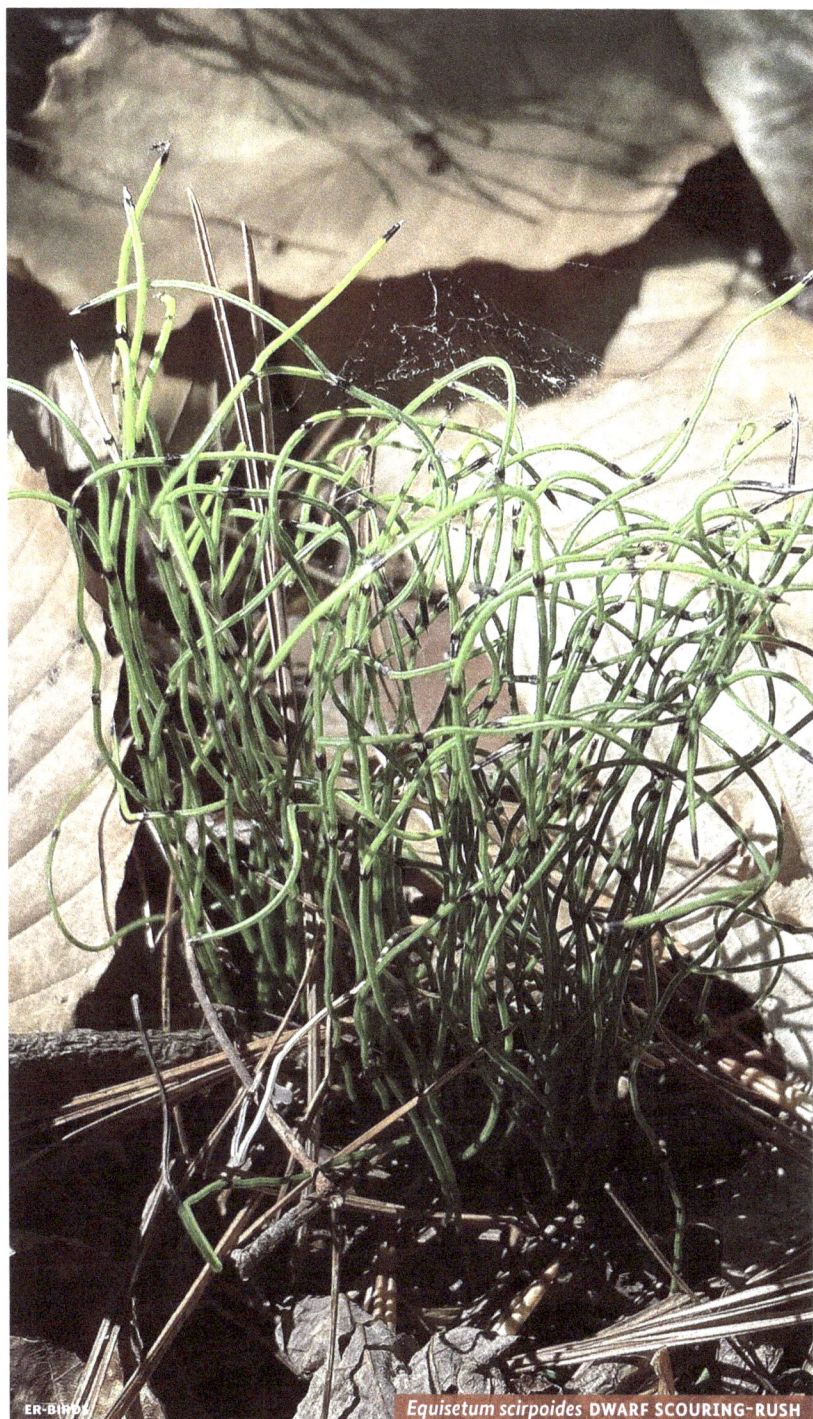

Equisetum scirpoides DWARF SCOURING-RUSH

Equisetum sylvaticum L.
WOODLAND HORSETAIL

DISTRIBUTION CT | MA | ME | NH | NJ | NY | PA | RI | VT

FIELD TIPS
- colony-forming horsetail of moist to wet woods
- stems deciduous, either sterile or fertile; sheaths loose and reddish brown.
- secondary branching of most primary branches
- cones rounded at tip

HABITAT

Swamps, moist forests and forest margins, wet thickets, moist meadows.

DESCRIPTION

ROOTSTOCK Light brown, shiny, occasionally tuber-bearing; forming colonies, the stems mostly solitary from the rhizome.

STEMS Annual, of two kinds, sterile and fertile.

STERILE STEMS To 70 cm long and 1.5–3 mm thick; central cavity 1/2 to 2/3 diameter of stem, with prominent outer cavities alternating with the 10–18 ridges; silica tubercules in 2 rows on the ridges; sheaths loosely inflated, the reddish brown papery teeth persistent, usually joined to form 3 or 4 groups; branches whorled, arched, appearing lacy due to the secondary branches; branches with 3 or 4 (rarely 5) ridges, and with narrow, pointed and spreading teeth.

FERTILE STEMS At first unbranched and lacking chlorophyll, fleshy, becoming green and branched after the spores released; sheaths and teeth usually larger than those of sterile stems.

CONES Stalked, to 3 cm in length, rounded at tip, shedding spores in spring then falling.

NAME Named by Linnaeus from European specimens in 1753.

FROM Latin, *sylva,* woods or forest, referring to its habitat.

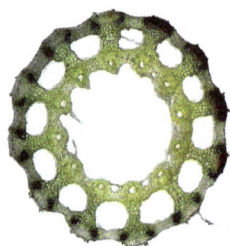

stem cross-section

BERND HAYNOLD

S. STIERLITZ

WOLFGANG JACKY

Equisetum sylvaticum **WOODLAND HORSETAIL**

Equisetum variegatum Schleich. ex F. Weber & D.M.H. Mohr
VARIEGATED SCOURING-RUSH

DISTRIBUTION CT | MA | ME | NH | NJ | NY | PA | RI | VT
FIELD TIPS
- colony-forming in moist, sandy, open places
- stems evergreen, slender, unbranched
- minutely grooved sheaths have 3–12 teeth with conspicuous white margins
- cones tipped by a small point

HABITAT
Moist sand, shores, beaches, interdunal flats, roadside ditches, borrow pits; often forming dense colonies.

DESCRIPTION
ROOTSTOCK Smooth, blackish, shiny, branching; the stems in clusters from the rhizomes.
STEMS Evergreen, 10–50 cm long, 0.5–3 mm thick, usually unbranched, ascending; central cavity 1/3 to 2/3 diameter of stem; large outer cavities alternating with 3–12 ridges; silica tubercules in two lines on ridges; sheaths green at base, black above, slightly spreading; teeth lance-shaped, persistent, with or without thread-like tips, and with a brown central portion and wide white margins.
CONES Small, 5–10 mm long, tipped by a small sharp point, shedding spores in late summer, or more often persisting unopened until the following spring.

stem cross-section

SYNONYMS *Hippochaete variegata* (Schleich. ex F. Weber & D. Mohr) Bruhin
CONSERVATION STATUS *endangered* New Jersey, Pennsylvania.
NAME This plant was discovered in Europe and named by Schleicher in 1807; American plants found soon afterward proved so similar that they are regarded as representing the same species. From Latin, *variego,* to be different colors.

GUILLAUME HOFFMANN

Equisetum variegatum VARIEGATED SCOURING-RUSH

HYMENOPHYLLACEAE *Filmy fern family*

Crepidomanes K. Presl **WEFT FERN**

WORLDWIDE, THE FILMY FERN FAMILY contains 9 genera and an estimated 434 species, mostly of tropical regions; *Crepidomanes* includes about 30 species, with a single species in North America. Weft ferns are very small, delicate, algae-like ferns known only in the gametophyte phase. Habitats are shaded crevices and hollows in noncalcareous rock; plants form small, matted colonies, usually on the ceiling of the hollow.

KEY CHARACTERS
- Tiny, thread-like fern fern of dark, moist rock crevices, caves, and hollows.
- Fronds translucent, 1–celled thick, adapted to low-light conditions.
- Forming small, matted colonies, plants resembling algae.
- Occurring only in the gametophyte phase of the fern life cycle.
- Sporangia in a cluster on a marginal bristlelike extension of a vein, surrounded at base by a cone-shaped sheath.

NOTE The region's smallest and most delicate fern; found only in the gametophyte stage.

NAME The fern forms felt-like mats, leading to the common name of 'weft fern' (weft is a term used in weaving). Our species formerly included in the larger genus *Trichomanes*.

Crepidomanes intricatum (Farrar) Ebihara & Weakley
WEFT FERN

DISTRIBUTION CT | MA | NH *historical* | NJ | NY | PA | VT
FIELD TIPS
- delicate evergreen fern of dark, moist, rock crevices and caves
- fronds translucent, 1–celled thick
- plants matted and forming colonies
- known only from the gametophyte phase

HABITAT

Shaded, moist, humid crevices and cave-like hollows in acidic sandstone and gneiss rock; plants often hanging from rock ceiling.

DESCRIPTION

Plants small and thread-like, forming small mats (to about 1 square meter) of intertwined stems; known only from the gametophyte phase; spore-bearing plants unknown in nature and have not been produced when cultured. Distinguished from algae by presence of short roots, gemmifers (flask-shaped cells) and tiny, thread-like gemmae (asexual propagules) produced by the gemmifers (these features visible under magnification).

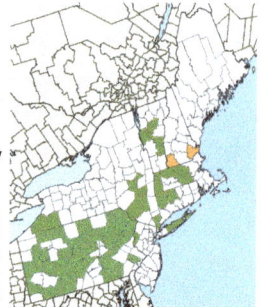

SYNONYMS *Trichomanes intricatum* Farrar
CONSERVATION STATUS *endangered* New Hampshire
(historical records only).

Crepidomanes intricatum WEFT FERN

LYGODIACEAE *Climbing fern family*

Lygodium Sw. CLIMBING FERN

FRONDS VINELIKE, OF INDETERMINATE GROWTH, pinnae reduced to short stalks, each bearing a pair of opposite pinnules, sporangia in 2 rows, 1 on each side of midvein of contracted, oblong, marginal lobes of ultimate segments, covered by hoodlike flap of tissue serving as the indusium.

Lygodium is the sole genus of this family; the genus includes an estimated 40 species, found mostly in tropical regions. North America is home to a single native species—*Lygodium palmatum*—and 2 species which have escaped from cultivation (below). *Lygodium* are a true anomaly within the fern world; they are climbing, vine-like ferns exhibiting indeterminate growth in which the rachis can reach 3–10 meters in length. Previously sometimes placed in Schizeaceae, the Curly-Grass Fern Family.

KEY CHARACTERS
• The region's only twining and climbing, vine-like ferns.

NAME From Greek, *lygodes,* flexible, referring to the rachis bending alternately from side to side.

ADDITIONAL MEMBERS OF GENUS
The two species below are introduced in North America. Japanese climbing fern is reported from western Pennsylvania. In the USA, *Lygodium microphyllum* is known only from Florida.

• **Japanese climbing fern [***Lygodium japonicum*** (Thunb. ex Murr.) Sw.]**, originally introduced to the USA as an ornamental plant, is established in the se USA, ranging from eastern Texas to North Carolina, with a disjunct occurrence reported from Beaver County in western Pennsylvania.

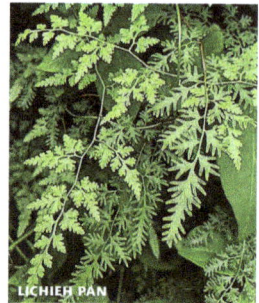
Lygodium japonicum

• **Old World climbing fern [***Lygodium microphyllum*** (Cav.) R. Br.]**, native to Australia, southern Asia and Africa, was first found in Florida in the 1960s and is now established on an estimated 25,000 hectares (about 100,000 acres) in central and southern Florida. An aggresive, kudzu-like weed, it infests cypress swamps, engulfing tree islands with fronds up to 30 meters long. Due to the warm climate, plants do not die back in winter, allowing for luxuriant growth. Efforts are ongoing to control the spread of this plant using biological agents including several moths, a beetle, and a mite.

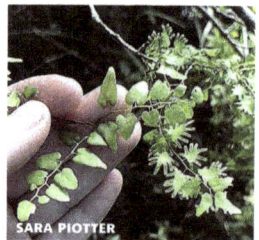
Lygodium microphyllum

Lygodium palmatum (Bernh.) Sw.
AMERICAN CLIMBING FERN

DISTRIBUTION CT | MA | NH | NJ | NY | PA | RI | VT *historical*

FIELD TIPS
- region's only climbing fern
- rachis vinelike, to 1 m or more long
- leaf blades sterile or fertile
- sterile blades 4–7-lobed
- fertile blades small, in terminal panicle

HABITAT

Woodland margins, wet thickets, in sun to light shade; soils moist and acidic.

DESCRIPTION

ROOTSTOCK Creeping, branched, slender, black, with sparse short, blackish hairs; scales absent.

FROND Divided into sterile and fertile portions, usually 40–100 cm long; climbing, forming apparently a "climbing stem" with palmately lobed "leaves" but actually a twining rachis bearing the pinnae.

RACHIS Twining, 2-forked, each fork with a palmately 4–7-lobed blade 2–7 cm wide; round and dark at base, upwards green or brownish and somewhat flattened and winged.

BLADE 2-pinnate, fertile blades uppermost, contracted and several times forked, forming a terminal panicle.

STERILE PINNAE Evergreen until following year's growth, alternate, each divided into two stalked pinnules.

FERTILE PINNAE Deciduous, above the sterile pinnae at tips of vine, several times branched.

SORI In a double row on fertile segments, under an indusiumlike flap of tissue at lobe edges of the smaller fertile pinnae.

CONSERVATION STATUS *endangered* New Hampshire, New York, Vermont; more common in southeastern USA.

NAME First found by William Bartram in Georgia, but he failed to give it a valid name. Plants from e USA collected by Muhlenberg were named *Hydroglossum* by Willdenow, and *Gisopteris palmata* by Bernhardi in 1801. Swartz transferred the species to his genus *Lygodium* 5 years later. Also known as Hartford fern, referring to its having been the subject of the first conservation law applied to a fern, passed by Connecticut legislature in 1869.

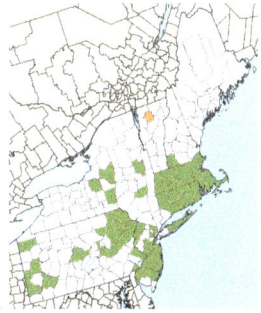

FROM Latin, *palmatum,* referring to the palmate shape of
the leaf blade.

DOUG GOLDMAN, USDA NRCS

DOUG MCGRADY

Lygodium palmatum AMERICAN CLIMBING FERN

MARSILEACEAE *Water-clover family*

Marsilea L. WATER-CLOVER

Perennial aquatic ferns with leaf blade divided into 4-segments. Family includes 3 genera and an estimated 55 species of *Marsilea,* mostly of tropical regions; one species in our flora, introduced from Europe and Asia.

KEY CHARACTERS

- Aquatic ferns of quiet, fresh water or sometimes exposed on shores; roots and often rhizomes anchored in bottom muck.
- Leaf blade 4-parted like a 4-leaf clover, atop a slender stipe of variable length, (length depending on depth of water).
- Pinnae floating on water surface or raised slightly above it, sometimes submerged.
- Spores of 2 kinds, the female, egg-bearing megaspores, and the male, sperm-bearing microspores; in 1–several hard, bean- or pea-like structures (sporocarps) near base of stipes.
- Sporocarps divided into several compartments, each surrounded by a gelatinous ring which serves to hold the megaspore until fertilization occurs by the microspore.

NAME After Count Luigi Fernando Marsigli (1658–1730), an Italian naturalist.

Marsilea quadrifolia L.

EUROPEAN WATER-CLOVER

DISTRIBUTION CT | MA | ME | NH | NJ | NY | PA

FIELD TIPS

- deciduous aquatic fern
- pinnae 4-parted, emergent to slightly below water surface
- sporocarps 1–2 on short stalks from near base of stipe

HABITAT

In mud or shallow water of quiet lakes, ponds and rivers.

DESCRIPTION

RHIZOME Creeping and rooted in bottom of lakes and ponds, less than 1 mm in diameter, smooth or with a few hairs; scales absent; roots forming both at nodes and sparsely (1–3) along internodes.

FROND Deciduous, sterile and fertile fronds alike,

STIPE Green, slender, to about 15 cm long,depending on water-depth.

BLADE Floating, submerged or emergent, horizontal, divided into 4 clover-like pinnae; smooth.

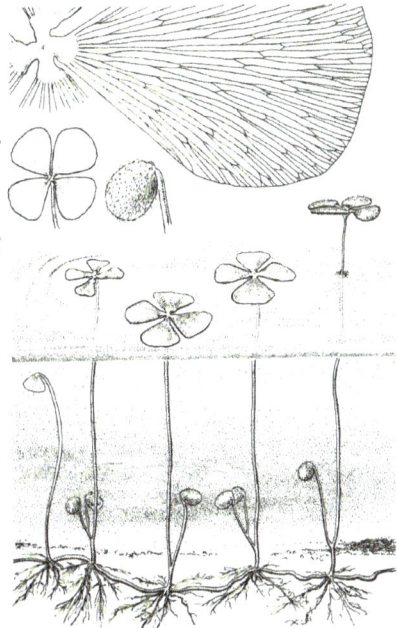

PINNAE 1.5–2.5 cm wide, smooth or with a few hairs; margins rounded; veins forked.

SPOROCARP Usually 2 on stalks from lower portion of stipe, shaped like a small bean, 2–5 mm long, with yellow hairs when young, hairless when mature.

NOTE Introduced to the USA from Europe in 1862, now naturalized and sporadic in eastern and midwestern states.

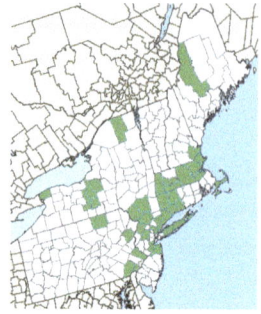

KRZYSZTOF ZIARNEK

CHRISTIAN BERG
Marsilea quadrifolia EUROPEAN WATER-CLOVER

ONOCLEACEAE *Sensitive fern family*

LARGE, COARSE FERNS with creeping hairy rhizomes (*Onoclea*) or with stolons on ground surface (*Matteuccia*); sterile and fertile fronds strongly different, the sterile fronds deciduous, pinnatifid to 1-pinnate-pinnatifid; fertile fronds persistent. Sori enclosed under recurved margin of pinna segment (outer false indusium) and a tiny true inner indusium (membranous or of hairs). Small family of four genera and an estimated five species; we have two genera, each with a single species.

KEY CHARACTERS
- Colony-forming ferns of moist to wet places.
- Sterile and fertile fronds strongly dissimilar
- Stipes with 2 vascular bundles, uniting upwards to form U-shape.
- Sori in beadlike clusters (*Onoclea*), or under rolled margins of fertile pinnae (*Matteuccia*).

KEY TO ONOCLEACEAE | SENSITIVE FERN FAMILY

1   Sterile blades solitary from creeping rhizomes, deeply divided into lobes (or the lowermost divisions pinnae); common, regionwide.... *Onoclea sensibilis*, SENSITIVE FERN, page 162
1   Sterile blades in a circle from a thick crown; pinnate with lobed pinnules; regionwide ... . *Matteuccia struthiopteris*, OSTRICH FERN, page 160

Matteuccia Todaro OSTRICH FERN

ONE OF THE REGION'S LARGEST FERNS, forming round tufts of bright green fronds. One circumboreal species in genus [but American and European populations perhaps best treated as distinct species (Koenemann et al., 2011)].

KEY CHARACTERS
- Large fern forming vase-shaped clumps, may form colonies via creeping stolons.
- Sterile and fertile fronds strongly different.
- Sterile frond 1-pinnate-pinnatifid, shaped like ostrich feather in outline, long-tapering to base.
- Fertile fronds shorter than sterile, turning dark brown, persistent into the following year; sori covered by inrolled margin of pinnae.

NAME
Named in honor of Carlo Matteucci (1811–1868), an Italian physicist.

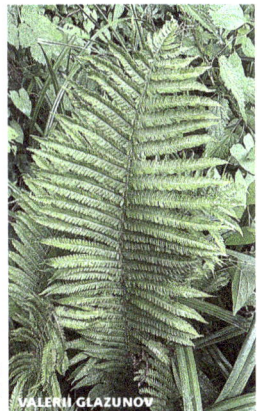

Matteuccia struthiopteris
OSTRICH FERN
VALERII GLAZUNOV

Matteuccia struthiopteris (L.) Todaro
OSTRICH FERN

DISTRIBUTION CT | MA | ME | NH | NJ | NY | PA | RI | VT

FIELD TIPS
- large, colony-forming, vase-like fern of wet places
- sterile fronds arched, widest in upper portion, lowest pinnae much smaller than upper
- fertile fronds upright from center of cluster, persistent

HABITAT

Streambanks, wet ravines, seeps and springs, wet woodlands, and swamps, soils typically alluvial, mucky or sandy; may form large colonies via creeping rhizomes.

DESCRIPTION

ROOTSTOCK Erect, stout, densely scaly, forming a vase-like cluster; also with spreading runners which produce several new plants each year.

CROZIER Green with deciduous tan scales; edible.

FROND Several in a cluster; sterile and fertile fronds very different; sterile fronds deciduous, 50–100 cm long, pinnate-pinnatifid. Fertile fronds 15–30 cm long, pinnate, dark green to blackish, drying to brown, often persist to next year.

STIPE Green, with white hairs; vascular bundles 2.

RACHIS Green, deeply grooved, with sparse whitish hairs, scales, or both.

STERILE BLADE Elliptic, widest in upper part, 1-pinnate-pinnatifid.

FERTILE BLADE 1-pinnate, dark green to blackish, drying to brown.

PINNAE Lowest pinnae (several pairs) greatly reduced in size; sessile; costae shallowly grooved above, grooves not continuous from rachis to costae; veins free, not forked, extending to margin.

SORI On margins, covered by inrolled edge of fertile pinnae; indusium vestigial; sporangia green, persist through winter and released in spring before sterile leaves expand.

SYNONYMS *Matteuccia pensylvanica* (Willd.) Raymond, *Onoclea struthiopteris* (L.) Hoffm. p.p., *Pteretis nodulosa* (Michx.) Nieuwl., *Pteretis pensylvanica* (Willd.) Fernald

SIMILAR SPECIES Fronds of **cinnamon fern** (*Osmundastrum cinnamomeum* and **interrupted fern** (*Osmundastrum claytonianum*) similar but do not have very small

pinnae at base of frond.

NOTE Croziers (fiddleheads) edible and prepared like aspara-
gus; sometimes harvested commercially (especially in New
England).

NAME In 1803 Michaux described *Onoclea nodulosa,* mistakenly ascribing it to Carolina. The identity of this fern remained a mystery for some time, but his type specimen
proved to be ostrich fern, and to have come from Montreal. From Greek, *struthos,* ostrich, and *pteris,* fern.

Matteuccia struthiopteris OSTRICH FERN

Onoclea L. SENSITIVE FERN

GENUS INCLUDES ONE SPECIES of eastern North America and eastern Asia.

KEY CHARACTERS
- Coarse fern of moist to wet places, forming colonies from creeping rhizome.
- Sterile and fertile fronds strongly different.
- Sterile fronds arising singly from rhizome, mostly pinnatifid.
- Fertile fronds turning dark brown; sori enclosed by the inrolled margins to form beadlike covering.

NAME From Greek *onos,* vessel, and *kleio,* to close, referring to closely rolled fertile fronds.

Onoclea sensibilis L.
SENSITIVE FERN

DISTRIBUTION CT | MA | ME | NH | NJ | NY | PA | RI | VT
FIELD TIPS
- coarse fern, may form large patches
- sterile fronds deciduous, triangular, not cut to rachis
- fertile fronds brown, with bead-like segments enclosing sori, persistent

HABITAT

Swamps, streambanks, thickets, marshes, moist woods; in sun or shade, often forming dense colonies.

DESCRIPTION

ROOTSTOCK Long-creeping, branched, producing a fibrous mat near the soil surface, smooth or with a few scales.

CROZIER Pale red.

FROND Sterile and fertile fronds different, arising singly from points along the rhizome; sterile fronds deciduous, to about 60 cm long, often tilted up and back, turning brown in late-sumer to early fall, even before frost. Fertile fronds shorter than sterile fronds, not evergreen but persistent and erect overwinter and sometimes for several years.

STIPE Usually longer than blade, swollen and dark brown at base, with a few light-brown scales, these deciduous; vascular bundles 2.

RACHIS Winged (sterile fronds), especially upward; smooth.

STERILE BLADES Broadly triangular, to 40 cm long and about as wide; 1-pinnate-pinnatifid at base, pinnatifid above.

FERTILE BLADES 2-pinnate, dark brown with buff rachises.

PINNAE 5 to 11 pairs, lowest pinnae largest or nearly so, stemless or attached laterally to rachis; upper pinnae connected to rachis by winglike tissue; fertile pinnules rolled into bead-like segments, enclosing the sporangia; margins entire to wavy; veins of sterile pinnae netted.

SORI Round, covered by rolled-up margins; indusium vestigial, not seen; sporangia green, the enveloping fertile pinnules becoming black in maturity; spores green in maturity, persist through winter and released in spring before sterile leaves expand.

NAME John Clayton found this fern in Virginia in the early 1700s, and named by Linnaeus in 1753. From Latin, *sensibilis,* sensitive, in reference to the sterile fronds quickly turning brown in late-summer or fall.

SANDY WOLKENBERG

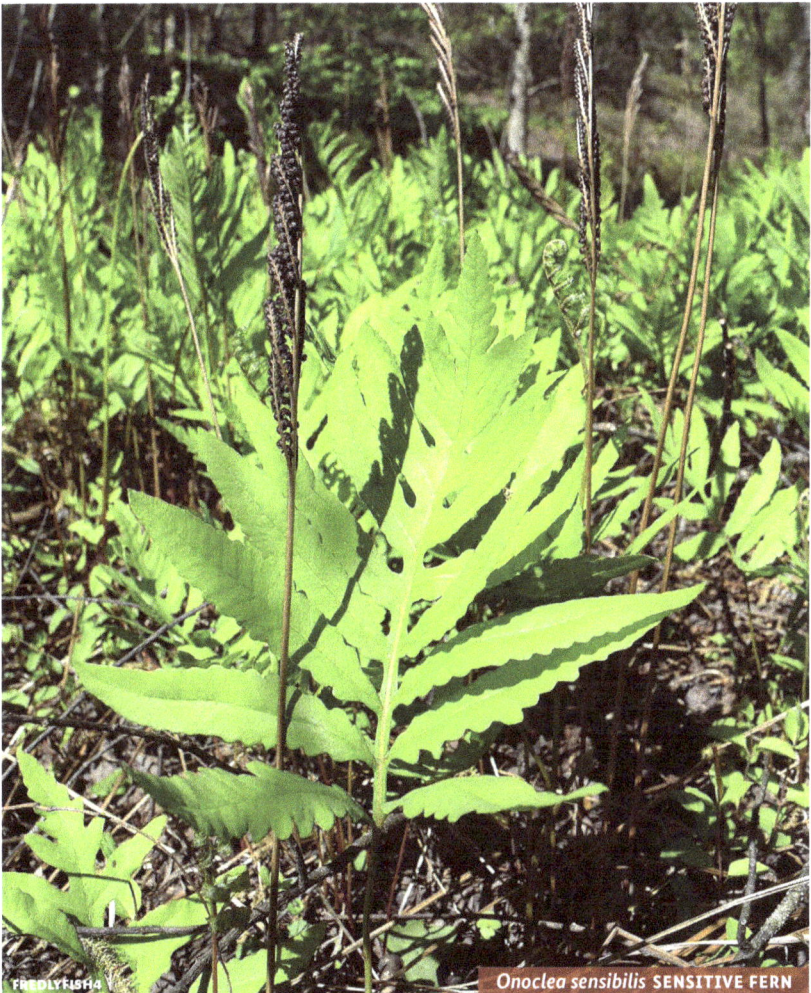

FREDLYFISH4

Onoclea sensibilis **SENSITIVE FERN**

OPHIOGLOSSACEAE *Adder's-tongue family*

SMALL TO MEDIUM, PERENNIAL, SOMEWHAT SUCCULENT FERNS. Roots few, fleshy, bearing a common stalk divided into a sterile blade portion (trophophore) and a fertile sporangia-bearing portion (sporophore); blade simple, divided, or compound; fertile portion without blade-like tissue, typically composed of a long stalk with a terminal, branched or unbranched, sporangia-bearing area; sporangia sphere-shaped, large (in comparison with those of most other ferns), about 1 mm in diameter, thick-walled, in 2 rows on the branches or on the unbranched sporangia-bearing stalk. The sphere-shaped sporangia are not covered by an indusium, lack an annulus or ring, and when mature, open via a transverse slit to release spores.

Unlike most ferns, new fronds do not emerge in the familiar fiddlehead form (circinate vernation), instead, the new shoot emerges uncoiled and straight in *Botrychium,* and somewhat bent in rattlesnake fern (Botrypus) and grape fern (*Sceptridium*). Gametophyte located underground, usually without chlorophyll and associated with mycorrhizal fungi.

In contrast to most true ferns (the "leptosporangiate" ferns), members of this family are part of the "eusporangiate" group which produce large, thick-walled sporangia, and perhaps only distantly related to other true ferns. Overall, a cosmopolitan family of 10 genera and an estimated 112 species, subdivided into four subfamilies, two in our flora: Botrychioideae (including genera *Botrychium, Botrypus, Sceptridium*) and Ophioglossoideae (genus *Ophioglossum*); these also sometimes recognized as distinct families.

KEY CHARACTERS
- Mostly small ferns, rhizomes and stalks fleshy.
- New shoots not coiled like a fiddlehead, instead emerging straight from the ground.
- Fronds divided into a single sterile blade (trophophore) and a spike-like or panicle-like fertile stalk (sporophore).
- Sporangia large, thick-walled.
- Habitats diverse, from open to shaded, moist to dry, sometimes where somewhat disturbed so that the litter layer is not very deep.

NOTE Taxonomically, rattlesnake fern (*Botrypus*) and grape ferns (*Sceptridium*) previously included in *Botrychium;* however, given the large number of moonworts and readily apparent differences amongst species, modern classifications separate the family into 3 subgenera or 3 distinct genera (as incorporated here).

NAME Family name from Greek, *ophis,* snake, *glossa,* tongue; the narrow fertile segment suggesting the name adder's-tongue.

Botrychium Sw. MOONWORT

Small, smooth ferns with fleshy stems, a single sterile blade (trophophore) and a
fertile blade (sporophore). Plants typically produce only one leaf each year from
an underground rhizome. Sterile blade divided or lobed, ovate to triangular in
outline; fertile portion of leaf consisting of an elongate stalk terminated by a
sporangia-bearing region.

Moonworts have a close association with mycorrhizal fungi on which they de-
pend for their water, mineral, and carbohydrate needs. *Botrychium* distribution
is probably highly related to the occurrence of these underground fungi.

A nearly cosmopolitan genus of approx. 50–60 species, with greatest diversity
at high latitudes and high elevations; about 30 species in North America; 8 in our
flora, some very rare.

KEY CHARACTERS

* **Small, somewhat fleshy, deciduous ferns.**
* **Fronds divided into a lobed or divided sterile blade and a fertile portion bearing
sporangia on its surface.**
* **Sporangia relatively large, often yellowish when mature.**

NOTE Most species quite variable in their form, and more than one species may occur in
suitable habitats, resulting in considerable taxonomic confusion; use the key, species
descriptions, habitat information, and distribution maps to confirm your identification. If collecting plants, cut rather than pinch the frond at ground level as underground parts are not necessary and are important to leave in the soil for future growth.

NAME From Greek, *botrys,* a cluster, referring to the clusters of sporangia. The common
name is in reference to the crescent-moon shape of the pinnae of the type species,
Botrychium lunaria, and the Old English 'wyrt', herb, for alleged medicinal properties.
Botrychium lunaria was first described in 1542 by Fuchs as *Lunaria minor.* Linnaeus
recognized two species of *Botrychium* in his 1753 Species Plantarum, *B. lunaria* and *B.*

virginiana. He placed both in the genus *Osmunda*. Presl (1845) was the first to use the name *Botrychium,* recognizing 17 species in his treatment of the genus. The first modern comprehensive treatment of the family was that of Clausen in his 1938 Monograph of the Ophioglossaceae.

MOONWORT FORMS

The shape of the trophophore (sterile portion) is the best diagnostic character for *Botrychium;* our species have one of three basic forms (Farrar, 2006):

A blade once-dissected (1-pinnate), pinnae fan-shaped;

B blade triangular, twice-dissected (2-pinnate);

C blade intermediate (pinnate-pinnatifid), form derived from ancestral hybridization between *B. angustisegmentum* and species of the fan-leaflet group.

Forms B and C are sometimes referred to as the **midribbed species** because their pinnae have strong central veins, whereas those of form A (pinnae fan-shaped) have multiple parallel veins of equal size. Presence of a midrib in the basal pinnae is a good way to identify plants in forms B and C when they are too small to have developed pinna lobing.

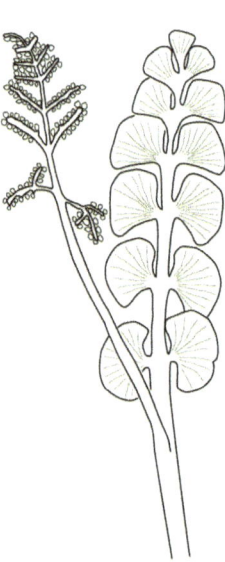

GROUP A	GROUP B	GROUP C
BLADE 1-PINNATE, PINNAE FAN-SHAPED	**BLADE TRIANGULAR 2-PINNATE**	**BLADE 1-PINNATE-PINNATIFID**
Botrychium ascendens	*B. angustsegmentum*	*B. matricariifolium*
B. campestre		
B. minganense		
B. neolunaria		
B. pallidum		
B. simplex		

KEY TO *BOTRYCHIUM* | MOONWORT

1 Sterile blade (trophophore) simple to lobed, lobes rounded to square and angular, stalks usually 1/2 to 2/3 length of sterile blade; plant often in open grassy fields; regionwide but absent from CT.................... *Botrychium simplex*, LEAST MOONWORT, page 176

1 Sterile blade pinnately lobed (either if actual pinnae or simply lobed), lobes of varying shapes, stalk usually less than 1/4 length of sterile blade; plants often in open sunny places; dunes, streambanks, roadsides, on trails and openings in forests, etc. 2

2 Basal pinnae or segments of sterile blade with venation like the ribs of a fan; midrib absent... 3

2 Basal pinnae or segments of sterile blade with pinnate venation; midrib present 6

3 Basal pinnae broadly fan-shaped (almost perfect half moons) with narrow stalks; rare in ME, NY..................... *Botrychium neolunaria*, COMMON MOONWORT, page 173

3 Basal pinnae narrowly fan- or wedge-shaped to nearly linear 4

4 Sterile blade flat or folded only at base when alive, usually up to 10 long by 2.5 cm wide; pinnae up to 10 pairs; basal pinnae unlobed, or if lobed, not usually 2-parted; ME, NY, VT .... *Botrychium minganense*, MINGAN MOONWORT, page 172

4 Sterile blade at least partially folded longitudinally when alive (conduplicate), usually not more than 4 cm long by 1 cm wide; pinnae up to 5 pairs; basal pinnae usually 2-parted. 5

5 Sterile blade very fleshy; fertile portion of blade usually less than 1.5 times length of sterile blade; pinnae mostly linear; basal pinna lobes ± equal; plants appearing in late spring; reported from southern VT .... *Botrychium campestre*, IOWA MOONWORT, page 170

5 Sterile blade herbaceous; fertile portion of blades usually 1.5–4 times the length of vegetative blades; pinnae asymmetrically fan-shaped; basal pinna lobes unequal; plants appearing in summer; rare in eastern ME *Botrychium pallidum*, PALE MOONWORT, page 175

6 Fertile portion of blade 3-parted, with 3 major branches from near base of stalk at sterile blade; regionwide.. *Botrychium angustisegmentum*, TRIANGULAR MOONWORT, page 168

6 Fertile portion of blade usually with loosely pinnate branches smaller than the single main stem; regionwide...... *Botrychium matricariifolium*, DAISY-LEAF MOONWORT, page 171

ADDITIONAL NORTHEASTERN SPECIES

Triangle-lobe moonwort (*Botrychium ascendens* W.H. Wagner), pinnae of sterile blade wedge-shaped, angled upward, outer margins toothed, lower pinnae sometimes divided into 2 lobes and also bearing sporangia; found in open, grassy fields; known in northeastern region only from Bennington and Windsor counties, Vermont.

Botrychium ascendens
TRIANGLE-LOBE MOONWORT

Botrychium angustisegmentum (Pease & A.H. Moore) Fernald
TRIANGULAR MOONWORT

DISTRIBUTION CT | MA | ME | NH | NJ | NY | PA | RI | VT

FIELD TIPS
• blade sessile, triangular, dark green and shiny
• pinnae deeply lobed

HABITAT

Most commonly in humus-rich leaf mold soils of moist de-ciduous and mixed conifer-hardwood forests, often with sugar maple (*Acer saccharum*).

DESCRIPTION

FROND 2–20 cm long or longer, fleshy.

STERILE BLADE Sessile, broadly triangular in outline, 2–3 cm long, dark green and shiny, from upper portion of common stalk; pinnae mostly 4–5 pairs, deeply lobed.

COMMON STALK Relatively long, brownish green.

SPOROPHORE Divided into several more or less equal, slightly spreading branches. Spores mature in July and August.

SYNONYM *Botrychium lanceolatum* subsp. *angustisegmentum* (Pease & A.H. Moore) R.T. Clausen

SIMILAR SPECIES **Daisy-leaf moonwort** (*Botrychium matricariifolium*) similar but in *B. angustisegmentum* sterile blade stalkless, basal pinnae much larger than other pinnae, and blade triangular in outline rather than trowel-shaped.

NOTE Formerly considered a subsp. of *Botrychium lanceolatum,* that species now considered to occur outside of our region in western North America.

NAME This is another of the species of grapeferns discovered by Gmelin in Russia, and assigned by him in 1768 to the genus *Osmunda;* transferred to *Botrychium* by Angström in 1854. From Latin, *lanceola,* a small lance, referring to the lance-shaped pinnae.

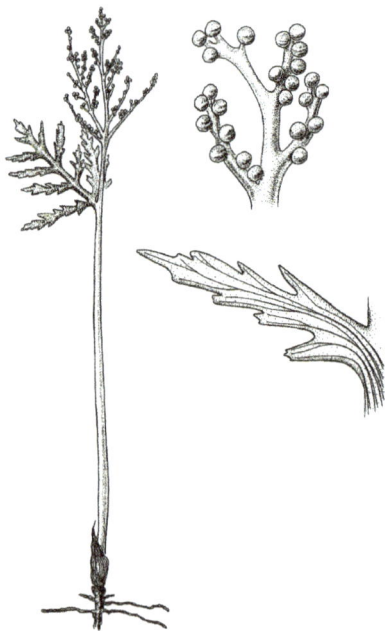

ER-BIRDS

Botrychium angustisegmentum TRIANGULAR MOONWORT

Botrychium campestre W.H. Wagner & Farrar
IOWA MOONWORT

DISTRIBUTION NY *historical* | VT

FIELD TIPS
• very small, inconspicuous fern of prairies, sand dunes and other sunny, calcareous sites
• stems and leaves succulent or fleshy
• southern Vermont only

HABITAT

Prairies, dunes, grasslands, and other open, calcareous sites.

DESCRIPTION

FROND Small, less than 5 cm tall and hard to find if surrounded by taller vegetation. Plants succulent, developing early in growing season, with aboveground portion drying by mid-summer.

STERILE BLADE Fleshy, once pinnate, sessile to the common stalk, oblong in outline, longitudinally folded; to about 4 cmlong and 1.3 cm wide. Usually divided into 5 pairs of linear or linear-spatulate segments; margins crenate or dentate and also usually notched into 2 or several smaller segments.

COMMON STALK Short, less than 3 cm long.

SPOROPHORE Often large relative to size of sterile blade.

CONSERVATION STATUS *endangered* New York (historical record from Onondaga County).

NOTES Tiny plantlets (gemmae) may be produced on underground stems.

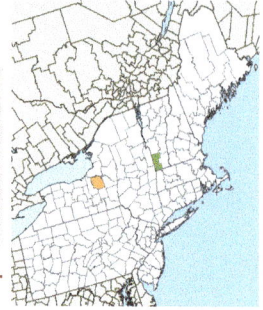

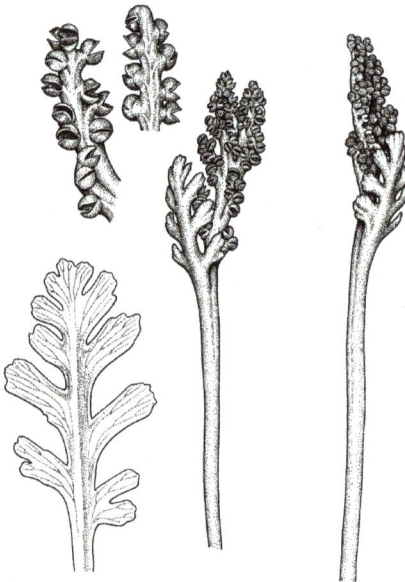

ER-BIRDS

Botrychium matricariifolium (A. Braun ex Dowell) A. Braun ex W.D.J. Koch

DAISY-LEAF MOONWORT

DISTRIBUTION CT | MA | ME | NH | NJ | NY | PA | RI | VT

FIELD TIPS

- plants to 20 cm tall
- sterile blade stalked, dull pale green, trowel-shaped in outline
- sporangia yellow

HABITAT

Most common on sandy, acidic soils such as old fields, clearings, and dry rocky woods; also reported from moist cedar woods or swamps.

DESCRIPTION

FRONDS Variable in shape and form, 5–20 cm long, somewhat fleshy.

STERILE BLADE Dull, pale green, trowel-shaped, longer than wide; varies from 1–3x pinnate; short-stalked, inserted above middle of common stalk; pinnae 2–7 pairs, the segments often deeply lobed.

COMMON STALK Slender, fleshy, pale green, often with a pinkish stripe.

SPOROPHORE Erect, usually with 3 main branches, these again divided; sporangia large, yellow.

SIMILAR SPECIES Triangular moonwort (*Botrychium angustisegmentum*), but in that species sterile blade nearly stalkless, and triangular rather than trowel-shaped in outline.

NAME Species name first suggested by A. Braun in 1843, and refers to similarity of blade to the dissected leaves of chamomile (*Matricaria*).

Botrychium minganense Victorin
MINGAN MOONWORT

DISTRIBUTION ME | NY | VT

FIELD TIPS
- plants to 20 cm tall but usually smaller
- sterile blade dull yellow-green, fleshy
- sporophore longer than sterile blade

HABITAT

Moist hardwood forests, aspen-balsam fir woods, and old clearings; soils mostly circumneutral.

DESCRIPTION

FRONDS Yellow-green, somewhat membranous, to about 20 cm tall, but usually much smaller.

STERILE BLADE Fleshy, dull yellowish green, stalked, emerges below or near middle of common stalk. Pinnae usually 6–10 pairs, ascending, fan-shaped or shallowly notched; sometimes with a few sporangia on margins of lower pinnae.

SPOROPHORE 1-pinnate, longer than sterile blade.

SYNONYM *Botrychium lunaria* subsp. *minganense* (Victorin) Calder & Roy L. Taylor

CONSERVATION STATUS *endangered* New York, Vermont (historical records only).

SIMILAR SPECIES
- Distinguished from **common moonwort** (*Botrychium neolunaria*) by its yellow-green color, stalked rather than sessile sterile blade, and narrower basal pinnae segments, which are not overlapping or as broadly fan-shaped.
- **Pale moonwort** (*Botrychium pallidum*) differs in being distinctly smaller and the pinnae whitish green.
- Somewhat similar to **spoon-leaf moonwort** (*B. spathulatum*) but the pinnae in that species narrower and spoon-shaped.

NAME Refers to Mingan Island in Gulf of St. Lawrence, where species first found as a new species (collected previously but confused with other species, especially *Botrychium simplex*). Earlier studies considered *Botrychium minganense* and *B. spathulatum* to be forms of the more common *B. neolunaria*.

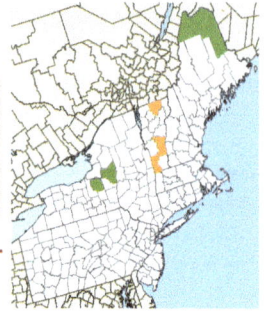

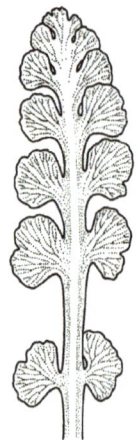

RYAN DURAND

Botrychium neolunaria Stensvold & Farrar
COMMON MOONWORT

DISTRIBUTION ME | NH | NY | VT *historical*

FIELD TIPS
- small moonwort of open habitats, often where calcium-rich
- sterile blade stalkless, shiny, fleshy, deep green
- pinnae segments fan-shaped

HABITAT

Open grassy fields and forests, usually on sandy or gravelly soil, sometimes over limestone and under trees of northern white cedar (*Thuja occidentalis*); also where somewhat disturbed.

DESCRIPTION

FRONDS Mostly less than 15 cm long, somewhat leathery.

STERILE BLADE Deep green, fleshy, shiny, oblong, to 10 cm long, stalkless or nearly so, inserted near or below middle of common stalk, 1-pinnate. Pinnae 6- 9 pairs, spreading, mostly overlapping except in shaded forest forms, basal pinna pair about equal in size and cutting to adjacent pair, broadly fan-shaped, undivided to tip, margins entire or undulate, rarely dentate, veins fanlike, midrib absent.

COMMON STALK 3–7 cm from base to juncture with sterile blade.

SPOROPHORE 1–2 pinnately branched, sometimes tip bending slightly downward; spores mature June-August.

SYNONYM *Botrychium onondagense* Underw. (now considered the shade form of *Botrychium neolunaria*).

CONSERVATION STATUS *endangered* Maine, New York.

SIMILAR SPECIES **Mingan moonwort** (*Botrychium minganense*) similar but sterile blade stalked and pinnae pairs separated and not overlapping. Most easily distinguished from other moonworts by spread of its pinnae; typically basal pinnae span an arc of nearly 180 degrees, 3rd pinna pair spans ca. 90 degrees.

NOTE Formerly named *Botrychium lunaria* (L.) Sw., that species common northward in Canada and Alaska (and reported from northeastern Maine). The subject of many tales in northern European folklore, including opening locked doors without a key; spores were said to make one invisible.

NAME From Latin, *luna,* moon, referring to the crescent-shaped pinna segments.

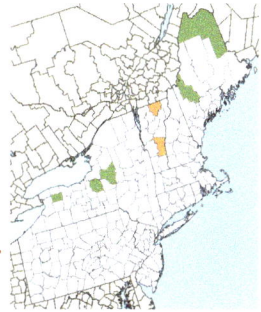

CHLOE & TREVOR VAN LOON

Botrychium neolunaria **COMMON MOONWORT**

Botrychium pallidum W.H. Wagner
PALE MOONWORT

DISTRIBUTION ME

FIELD TIPS
- plants very small, waxy pale green to whitish
- pinnae usually folded lengthwise
- lower pinnae often 2-lobed, sometimes with sporangia
- in our region, known only from eastern Maine

HABITAT
Open fields, dry sand and gravel ridges, roadsides, wet depressions, marshy lakeshores, tailings basins, second-growth forests; soils sandy.

DESCRIPTION

FROND Small, 2.5–7 cm long, waxy pale green to whitish.

STERILE BLADE To 4 cm long by 1 cm wide, 1-pinnate, with up to 5 pairs of fan-shaped pinnae, each pair of pinnae often folded towards each other; basal pinnae usually divided into 2 unequal lobes, the upper lobe larger; margins entire to irregularly toothed. Sporangia sometimes present on lobes of lower pinnae.

SPOROPHORE Longer than sterile blade, the sporangia on short branches from main stalk.

CONSERVATION STATUS *special concern*
Maine (Washington County).

SIMILAR SPECIES **Mingan moonwort** (*Botrychium minganense*) similar but larger and yellow-green rather than pale green as in *B. pallidum;* pinnae not folded as in *B. pallidum.*

NOTE Often with dense clusters of tiny, spherical gemmae (plantlets) on underground roots.

NATE MARTINEAU

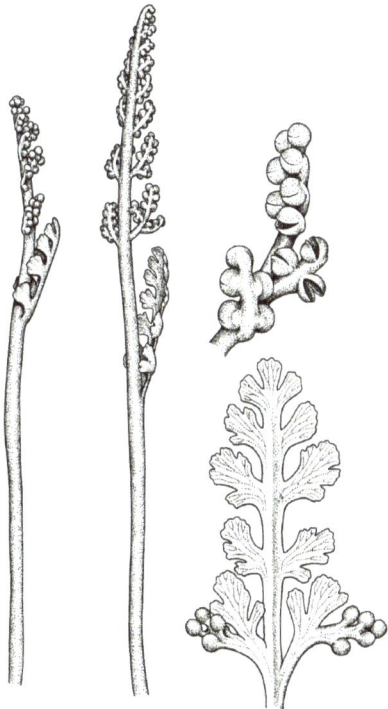

Botrychium simplex E. Hitchc.
LEAST MOONWORT

DISTRIBUTION MA | ME | NH | NY | PA | RI | VT | CT, NJ *historical*

FIELD TIPS
- small plants, dull to bright green to whitish green
- sterile blade simple to 1-pinnate; lowest pinnae usually largest, divided
- sporophore often unbranched

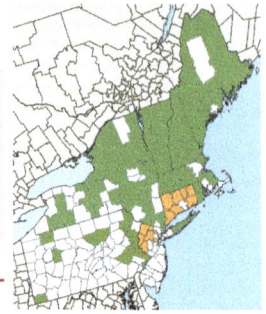

HABITAT

In a variety of open to shaded habitats: prairies, clearings, bracken fern grasslands, sandy woods, swamps and lakeshores. *Botrychium simplex* var. *tenebrosum* more common in forests, especially low places in moist hardwood forests or on mossy hummocks in conifer swamps.

DESCRIPTION

FROND Variable in size, to 15 cm long, often much smaller; rather fleshy; dull to bright green to whitish green.

STERILE BLADE Stalked; simple, lobed or pinnately divided with up to 7 pairs of well-developed, often overlapping lobes; small plants may have only 3 segments; basal pair of lobes usually much larger and more divided than upper pairs; margins nearly entire; sometimes with a few sporangia on margins of lower pinnae.

COMMON STALK blades attached at base or towards middle.

SPOROPHORE Simple or 1-pinnate, 1–8x length of sterile blade. Spores mature in late May and June.

SYNONYM *Botrychium tenebrosum* A.A. Eaton

SIMILAR SPECIES Very small plants may appear similar to **little goblin moonwort** (*Botrychium mormo*), a species of Quebec and the upper midwest and which has fleshier leaves and stalk.

NOTE Sterile portion variable, which has led to the naming of several taxonomic varieties, separated as follows:

- var. *simplex* (Lasch) R.T. Clausen: junction of sporophore and sterile blade near ground level, basal pinnae largest; stalk of sterile blade almost as long as blade.
- var. *tenebrosum* A.A. Eaton: junction of sporophore and sterile blade higher, well above ground level, basal pinnae more or less same size as upper pinnae; stalk of sterile blade much shorter than blade.

Sometimes considered as a distinct species: *Botrychium tenebrosum* A.A. Eat.

NAME Most circumboreal species have been named in Europe first, but the present one was overlooked there by the early workers, and described from Massachusetts by Hitchcock in 1823. *Simplex,* simple, may refer to the shallowly lobed sterile blade or to the sometimes unbranched sporophore.

MARTY PURDY

Botrychium simplex LEAST MOONWORT

Botrypus Michx. **RATTLESNAKE FERN**

A large and usually easily identified member of the Adder's-tongue Family. Plants appear early in spring, the sterile blade often persisting into fall, the fertile portion (sporophore) withering in early summer. Sterile blade broadly triangular and usually more than 10 cm wide, and sometimes as much as 30 cm wide. Fertile stalk attached directly below base of sterile blade, and both the blade and fertile branch are held well above the ground. Sterile plants can usually be recognized by the large size and the elevated blade. Worldwide, the genus includes 7 species, one in North America and in our flora.

KEY CHARACTERS
- Our largest and most common member of the Ophioglossaceae.
- Leaf blade nearly horizontal, broadly triangular in outline, highly dissected, bright green and shiny, deciduous.
- Sporophore long-stalked, overtopping sterile blade, branched, not persistent.
- Sporangia bright yellow.
- Usually in rich shaded woods, but also found in conifer swamps and drier deciduous forests.

NOTE *Botrypus* and the **grape ferns** (*Sceptridium*) are distinguished from **moonworts** (*Botrychium*) by their generally larger size, and separated from one another by different points of attachment of their sporophores: in rattlesnake fern (*Botrypus*), trophophore (sterile portion) and sporophore joined well above ground level, trophophore not stalked; in *Sceptridium,* trophophore and sporophore joined near ground level, trophophore stalked.

NAME From Greek, *botry,* bunch of grapes, from resemblance of the sporangia to clusters of grapes. Common name from reported use of the mashed roots by Native Americans to treat bites of poisonous snakes. The clusters of unopened sporangia are also somewhat similar to the rattles on tail of a rattlesnake.

Botrypus virginianus
RATTLESNAKE FERN

Botrypus virginianus (L.) Michx.
RATTLESNAKE FERN

DISTRIBUTION CT | MA | ME | NH | NJ | NY | PA | RI | VT
FIELD TIPS
- our most common member of family
- plants large, bright green, smooth
- sterile blade broadly triangular
- sporophore above sterile blade, not persisting through summer
- sporangia bright yellow

HABITAT
Most common in moist deciduous forests, also in drier woods and less commonly on hummocks in cedar swamps.

DESCRIPTION

FROND Deciduous, erect, smooth or nearly so, to about 50 cm long.

STERILE BLADE Broadly triangular, usually more than 10 cm wide, blades of larger plants to 30 cm wide; membranous or slightly fleshy, attached above middle of the stipe, 2-pinnate to 3-pinnate; margins toothed; veins few.

COMMON STALK Erect, smooth, fleshy, pinkish at base; fertile branch attached directly below base of sterile blade and both are held well above ground level.

SPOROPHORE Pinnately compound, produced in spring but withering in early summer; sporangia bright yellow.

SYNONYM *Botrychium virginianum* (L.) Sw.

SIMILAR SPECIES Small plants somewhat resemble **triangular moonwort** (*Botrychium angustisegmentum*), which also has glossy, triangular sterile blades. However, the sporophore of rattlesnake fern is long-stalked compared to the very short stalk of the sporophore of *B. angustisegmentum.* The large size, highly divided, nearly horizontal blade, and characteristic fertile segment make identification straight-forward.

NOTE The most common and widespread member of the family, found in every state except Hawaii, Nevada and Utah.

NAME Although also found in Europe and Asia, species first named based on specimens from Virginia by Linnaeus in 1753. His name for it was *Osmunda virginiana,* but transferred to genus *Botrychium* by Swartz in 1801. Now usually placed in own genus *Botrypus.*

JOHN KNOUSE

Botrypus virginianus RATTLESNAKE FERN

Ophioglossum L. ADDER'S-TONGUE

SMALL, DECIDUOUS FERNS with a single simple leaf and a stalked fertile portion. Worldwide, the genus has 41 reported species, mainly of tropical and subtropical regions; 3 species in our flora.

KEY CHARACTERS

• Small plants of moist to dry, usually open, grassy habitats.
• Leaf blade fleshy, simple, midrib absent, the veins netlike.
• Fertile portion an elongate stalk tipped by an unbranched, spikelike, sporangia-bearing region; the sporangia deeply embedded and in 2 rows.

NOTE *Ophioglossum* has the highest number of chromosomes known for vascular plants, with as many as 1,200 chromosomes in each cell reported.

Plants of *Ophioglossum* are small and easily overlooked, and also resemble sterile lilies or plantains (*Plantago*). However, *Ophioglossum* species can be distinguished by the leaf blade's netlike venation (not parallel) and lack of a midrib.

Leaf venation can be useful for identifying some *Ophioglossum* species. However, as this may be hard to see in dried specimens, Wagner et al. (1984) suggested wetting the leaf with several drops of 95% ethanol and then observing using transmitted light.

NAME From Greek, *ophis,* snake, *glossa,* tongue; the narrow fertile segment suggesting the name adder's-tongue.

Ophioglossum vulgatum
SOUTHERN ADDER'S-TONGUE

GAIL HAMPSHIRE

KEY TO *OPHIOGLOSSUM* | ADDER'S-TONGUE

1   Leaf blade abruptly narrowed a tip to a small sharp point; rare in southern PA . . . .
     . . . . *Ophioglossum engelmannii*, LIMESTONE ADDER'S-TONGUE, page 182
1   Tip of leaf blade blunt; more widely distributed in northeast region 2
2   Leaves gradually tapered to the base, elliptic in outline (widest near the middle); region-
     wide . . . . *Ophioglossum pusillum*, NORTHERN ADDER'S-TONGUE, page 183
2   Leaves abruptly tapered to the base, ovate in outline (clearly widest below the middle); CT,
     NJ, PA. . . . . *Ophioglossum vulgatum*, SOUTHERN ADDER'S-TONGUE, page 184

Ophioglossum engelmannii Prantl
LIMESTONE ADDER'S-TONGUE

DISTRIBUTION PA

FIELD TIPS
• restricted to limestone; rare in southern Pennsylvania
• plants pale green, with 1 (rarely 2) simple leaves
• leaf blade tipped with small sharp point

HABITAT
Dry limestone and dolomite barrens, glades and open woods.

DESCRIPTION

FROND Pale green, somewhat fleshy, 10–20 cm tall, sometimes producing 2 leaves; typically forming colonies.

STERILE BLADE Simple, entire, elliptical, sharp-pointed at tip, smooth; 2.5 to 3.5 cm long and 1 to 4.5 cm wide.

SPOROPHORE 5–10 cm long, the portion bearing sporangia to 5 cm; with 20–40 sporangia pairs, tapered at tip to a narrow point.

SYNONYM *Ophioglossum vulgatum* var. *engelmanii* (Prantl) Clute

CONSERVATION STATUS *endangered* Pennsylvania (Franklin County).

SIMILAR SPECIES Similar to **southern adder's-tongue** (*Ophioglossum vulgatum*), with overlapping range and sometimes sharing same habitat. However, plants of *O. engelmannii* have small, sharp points on the tip of blade, the upper half of the blades curve downward, and the main veins form large primary areoles that enclose several secondary areoles (as opposed to *O. vulgatum* which does not have secondary areoles). Fronds of limestone adder's-tongue are thick and almost succulent, and paler green than those of *O. vulgatum.*

NAME Sent from Missouri to Prantl by botanist George Engelmann, and named for the latter in 1883.

SAMANTHA HELLER

Ophioglossum pusillum Raf.
NORTHERN ADDER'S-TONGUE

DISTRIBUTION CT | MA | ME | NH | NJ | NY | PA | RI | VT

FIELD TIPS
- small plant of grassy, wet to moist places
- blade pale green, simple, widest near middle
- sporangia in 2 rows on fertile branch

HABITAT

Moist sandy fields, wet meadows, ditches, also in drier situations; soils typically sandy; not in dense shade.

DESCRIPTION

FROND 15–25 cm long , with a single, simple blade; fertile branch (sporophore) arising near base of blade.

STERILE BLADE Sessile, smooth, entire, attached near middle of stalk; broadly lance-shaped to ovate, broadest near the middle; 4–10 cm long, 1.5–3 cm wide.

SPOROPHORE Stalked spike extending 5–10 cm above blade, bearing 10–40 pairs of sporangia.

SYNONYMS *Ophioglossum vulgatum* auct. non L., *Ophioglossum vulgatum* var. *alaskanum* (E.G. Britton) C. Chr., *Ophioglossum vulgatum* var. *pseudopodum* (S.F. Blake) Farw.

CONSERVATION STATUS *endangered* New Hampshire, Rhode Island; *threatened* Connecticut, Massachusetts.

SIMILAR SPECIES **Southern adder's-tongue** (*Ophioglossum vulgatum*) similar but blade widest near base and darker green; in northern adder's-tongue, blade widest at middle and pale green.

NAME From Latin, *pusillus,* very small, perhaps alluding to its concealment by surrounding vegetation.

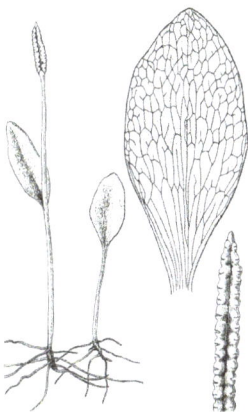

Ophioglossum pusillum **NORTHERN ADDER'S-TONGUE**

Ophioglossum vulgatum L.
SOUTHERN ADDER'S-TONGUE

DISTRIBUTION CT | NJ | PA

FIELD TIPS
- small fern of moist woods and fields
- sterile blade dark green, widest near base
- sporangia in 2 rows on fertile branch

HABITAT

Moist deciduous forests, moist meadows, thickets and clearings.

DESCRIPTION

FRONDS Scattered from rhizome, to about 25 cm long. divided into lower sterile and upper fertile portions; withering by midsummer.

STERILE BLADE Simple, oval to ovate, widest near the stalk-less base, rounded at tip, to 10 cm long and 4 cm wide, smooth and fleshy, dark green; veins netlike.

SPOROPHORE Stalked spike above sterile blade, with 10–35 pairs of sporangia; spores spherical, yellow.

SYNONYM *Ophioglossum pycnostichum* (Fernald) Á. & D. Löve

CONSERVATION STATUS *endangered* New Jersey.

SIMILAR SPECIES **Northern adder's-tongue** (*Ophioglossum pusillum*), but blade in that species is paler green and widest at middle.

NOTE Traditional European folk use of leaves and rhizomes as a poultice for wounds. This remedy was sometimes called the 'Green Oil of Charity.' A tea made from the leaves was used as a traditional European folk remedy for internal bleeding and vomiting.

NAME Named from European specimens by Linnaeus in 1753; American plants are identical with European plants. Epithet from Latin, *vulgaris,* common; the most common species in Europe and North America.

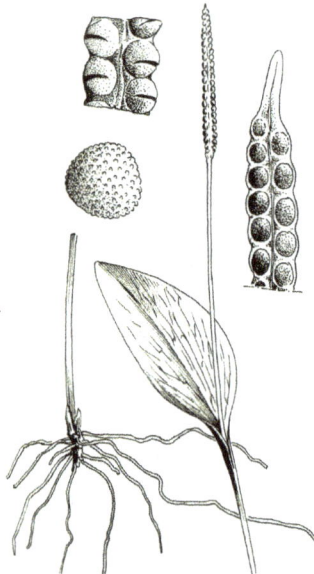

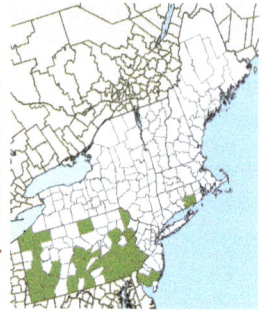

DMITRY IVANOV

Sceptridium Lyon GRAPE FERN

SMALL TO MEDIUM LEATHERY FERNS found in a variety of moist to dry, open to shaded habitats, often where sandy. Sterile blades dissected, winter-green, sporophore short-lived, withering by late summer. Sterile blade and sporophore joined near or below ground level. Worldwide, 10 species recognized, with 5 species in our flora.

KEY CHARACTERS
• Plants emerge in early summer, sterile blade persisting over winter.
• Sterile blade dissected, often appearing 3-parted; leathery or herbaceous.
• Sporophore short-lived, not overwintering; bearing yellow sporangia in a branched cluster

SIMILAR GENERA
Grape ferns (*Sceptridium*) distinguished from **moonworts** (*Botrychium*) by their generally larger size, usually leathery texture, the sterile blade nearly parallel to ground surface, and the joining of sterile blade and sporophore at or below ground surface.

   Sceptridium and *Botrypus* separated from one another by different points of attachment of their sporophores:
• In **rattlesnake fern** (*Botrypus*), sterile blade (trophophore) and sporophore joined well above ground level, sterile blade not stalked;
• In **grape ferns** (*Sceptridium*), sterile blade and sporophore joined near ground level, sterile blade stalked.

NAME From Greek, *sceptrum,* from resemblance of sporophore to a scepter or staff. Common name 'grape fern' from resemblance of sporangia to small clusters of grapes.

Sceptridium multifidum

KEY TO *SCEPTRIDIUM* | GRAPE FERN

1 Sterile blade segments deeply cut more than half way to the midvein, the entire blade lacerate 2

1 Sterile blade segments finely to coarsely toothed 3

2 Sterile blade mostly 2-pinnate, the segments sharply serrulate; rare in CT, NJ . . . .
. . . . *Sceptridium biternatum*, SPARSE-LOBE GRAPE FERN, page 187

2 Sterile blade mostly 3-pinnate (or more divided), the segments entire to obscurely serrulate or crenulate; common, regionwide . . . . *Sceptridium dissectum*
. . . . CUT-LEAF GRAPE FERN, page 188

3 Ultimate segments of blade ± uniform in size; sterile blade segments finely toothed to ± entire; dissection of blade into segments extending to within 1 cm of apex at tips of blades 4

3 Ultimate segments of blades variable in size, the apical segments much longer than the laterals; sterile blade segments coarsely and ± irregularly toothed or cut; dissection of blade into segments stopping at ca. 1–2.5 cm from apex at tips of blades 5

4   Segments of sterile blade rounded at base; symmetrically tapered to an often ± blunt or even rounded apex; larger segments mostly 9–17 mm long; margins nearly entire or finely and inconspicuously toothed; regionwide . . . . *Sceptridium multifidum*
. . . . LEATHERY GRAPE FERN, page 190

4   Segments of sterile blade usually (obliquely) asymmetrical and angular, cuneate to the apex; larger segments mostly 4–9 mm long; margins clearly finely dentate, especially visible in immature leaves; CT, NY, VT . *Sceptridium rugulosum*, TERNATE GRAPE FERN, page 194

5   Overwintering leaves green, not bronze; larger (terminal) segments of vegetative blades narrowly to broadly ovate, obtuse to rounded at apex, ± symmetrical at base; margins toothed but never lacerate; regionwide . . . . *Sceptridium oneidense*
. . . . BLUNT-LOBE GRAPE FERN, page 192

5   Overwintering leaves bronze-colored (or green if covered by leaves); larger (terminal) segments of sterile blades lance-shaped, acute, and strongly asymmetric at base; margins toothed to irregularly cut 6

6   Sterile blade mostly 2-pinnate, the segments sharply serrulate; CT, NJ . . . .
. . . . *Sceptridium biternatum*, SPARSE-LOBE GRAPE FERN, page 187

6   Sterile blade mostly 3-pinnate (or more divided, those forms keyed above), the segments entire to obscurely serrulate or crenulate; common, regionwide . . *Sceptridium dissectum*
. . . . CUT-LEAF GRAPE FERN, page 188

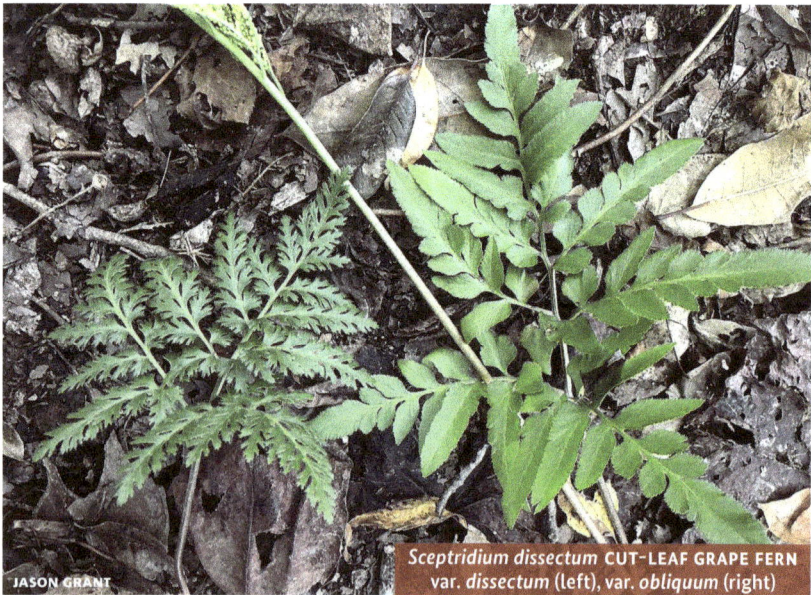

Sceptridium dissectum CUT-LEAF GRAPE FERN
var. *dissectum* (left), var. *obliquum* (right)

JASON GRANT

Sceptridium biternatum (Sav.) Lyon
SPARSE-LOBE GRAPE FERN

DISTRIBUTION CT | NJ

FIELD TIPS
- sterile blade evergreen, 3-parted, pinnae less divided than our other grape ferns
- uncommon in southern portions of region

HABITAT

A variety of moist, shaded situations; low woods, ravines, thickets, forested floodplains.

DESCRIPTION

STERILE BLADE Dark green (and remaining green over winter), herbaceous, horizontal; 3-parted, 1–2-pinnate; pinnules few, elongate, sparsely cleft; margins sharp-toothed; veins forked.

COMMON STALK Short, mostly underground.

SPOROPHORE Deciduous, several times longer than sterile blade, branched; sporangia yellow, in tight clusters.

SYNONYMS *Botrychium biternatum* (Sav.) Underw., *Botrychium dissectum* var. *tenuifolium* (Underw.) Farw.

NOTE Usually less dissected than **cut-leaf grape fern** (*Sceptridium dissectum*), and the pinnae trowel-shaped; more common in southeastern USA.

MICHELLE WONG

Sceptridium dissectum (Spreng.) Lyon
CUT-LEAF GRAPE FERN

DISTRIBUTION CT | MA | ME | NH | NJ | NY | PA | RI | VT

FIELD TIPS
- common grape fern of moist to dry, usually open and sandy places
- blade somewhat leathery, over-wintering
- blade color and dissection highly variable

HABITAT

Dry hilltops, dry to moist sandy fields, pastures and open woods; sometimes along old trails and roads.

DESCRIPTION

FROND Sterile blade diverges from sporophore at or below ground level; the single leaf is held close to the ground surface and overwinters, often turning a bronze or reddish color late in growing season following frosts.

STERILE BLADE Long-stalked, broadly triangular in outline, about 10–30 cm long and as wide, 3-parted; ultimate divisions of blade cut in linear segments.

SPOROPHORE Stalked, panicle-like, with clusters of light yellow sporangia; spores mature in fall.

SYNONYMS *Botrychium dissectum* Spreng., *Botrychium dissectum* var. *obliquum* (Muhl. ex Willd.) Clute, *Botrychium dissectum* var. *oblongifolium* (Graves) Broun, *Botrychium obliquum* Muhl. ex Willd.

SIMILAR SPECIES
- Leaves less leathery than the similar **leathery grape fern** (*Sceptridium multifidum*).
- **Rattlesnake fern** (*Botrypus virginianus*) has somewhat similar dissected blade but is larger, and the sterile blade and sporophore are attached well above the ground.

NOTE Extremely dissected and lacerated plants appear almost skeletonized and are very distinctive, and sometimes considered separate varieties (however, a wide range of forms may be present, even within a single population, see photo on page 186):
- var. **dissectum**: pinnules deeply cut and margins sharply toothed.
- var. **obliquum**: pinnules less deeply dissected, and may be confused with **blunt-lobe grape fern** (*Sceptridium oneidense*) which has blunter pinnules, and blunt or small teeth.

NAME This fern was discovered in Virginia and sent to Sprengel for identification, being named by him in 1804.

Sceptridium dissectum CUT-LEAF GRAPE FERN

STEPHEN

Sceptridium multifidum (Gmel.) Nishida ex Tagawa
LEATHERY GRAPE FERN

DISTRIBUTION CT | MA | ME | NH | NJ | NY | PA | RI *historical* | VT

FIELD TIPS
- region's largest grape fern
- blade leathery, succulent, winter-green, not turning bronze-colored in winter
- sporophore large, branched

HABITAT

Dry to moist sandy barrens, openings in oak and pine woods, old fields and clearings, along trails and old roads, grassy meadows and lawns; soils acidic; sterile plants sometimes found in shaded deciduous woods.

DESCRIPTION

FROND Up to 20 cm long, our largest *Sceptridium;* sterile blade and stalk very leathery and succulent; blade and sporophore attached near base of plant.

STERILE BLADE Long-stalked, broadly triangular in outline, 3-parted, shiny green to gray-green, evergreen, to 20 cm long and wide; held horizontally or bent backwards. Pinnae long-stalked, variable, ultimate segments crowded, sometimes overlapping.

SPOROPHORE (if present) large and panicle-like; sporangia numerous, yellow, maturing in late-summer.

SYNONYMS *Botrychium californicum* Underw., *Botrychium coulteri* Underw., *Botrychium matricariae* (Schrank) Spreng., *Botrychium multifidum* (S.G. Gmel.) Trevis., *Botrychium silaifolium* C. Presl, *Sceptridium silaifolium* (C. Presl) Lyon

CONSERVATION STATUS *endangered* New Jersey.

NOTE Previous year's blade remains green through winter until a new leaf appears in spring, then sometimes yellowing and withering but persisting through much of the following summer.

NAME In 1766 Gmelin applied the name *Osmunda multifida* to a Russian species; this changed to *Botrychium multifidum* by Ruprecht in 1859. The related plant occupying northern North America was named *B. silaifolium* by Presl in 1825, but is considered only varietally distinct from those in Europe. In 1916, Farwell proposed *Botrychium multifidum* for North American plants.

VALERII GLAZUNOV

ED ALVERSON

Sceptridium multifidum LEATHERY GRAPE FERN

Sceptridium oneidense (Gilbert) Holub
BLUNT-LOBE GRAPE FERN

DISTRIBUTION CT | MA | ME | NH | NJ | NY | PA | RI | VT

FIELD TIPS
- sterile blade bright or bluish green, remaining green through winter
- sporophore large, sporangia yellow
- moist to wet, shaded woods

HABITAT

Shaded, wet to dry, hardwood or mixed forests, occasionally in conifer swamps.

DESCRIPTION

FROND To 40 cm long or longer; stem and sterile blade somewhat leathery; sterile blade diverges from the common stalk about 2.5–5 cm above the ground.

STERILE BLADE Winter-green, bluish green, more or less flat, broadly triangular in outline, 3-parted, to 20 cm long and 15 cm wide, borne near the ground; pinnules blunt at tip, margins with fine teeth.

SPOROPHORE Panicle-like, not always produced each year; sporangia yellow, maturing in fall.

SYNONYMS *Botrychium oneidense* (Gilbert) House, *Botrychium dissectum* var. *oneidense* (Gilbert) Farw., *Botrychium multifidum* var. *oneidense* (Gilbert) Farw.

CONSERVATION STATUS *endangered* New York.

SIMILAR SPECIES
- **Leathery grape fern** (*Sceptridium multifidum*) similar in size and shape, but blunt-lobe grape fern has more unequally divided pinnae segments (the upper segments longer and less divided than the basal segments), and margins usually somewhat fine- toothed. In *S. multifidum*, segments usually wavy and lobed.
- **Cut-leaf grape fern** (*Sceptridium dissectum*) similar, but pinnae segments of blunt-lobe grape fern are blunt tipped, whereas those of *S. dissectum* somewhat pointed; *S. oneidense* leaves mostly green, those of cut-leaf grape fern reddish when young, and often bronzy in winter.

NOTE Previously treated as a variety, form, or hybrid of both *Sceptridium dissectum* and *S. multifidum*. The broader, more rounded divisions and the more shaded habitats of blunt-lobe grape fern are characteristic.

MATT BERGER

NAME 'Oneidense' refers to Oneida County, New York.

Sceptridium oneidense BLUNT-LOBE GRAPE FERN

Sceptridium rugulosum (W.H. Wagner) Skoda & Holub
TERNATE GRAPE FERN

DISTRIBUTION CT | NY | VT

FIELD TIPS
• sterile blade herbaceous, not leathery
• blade segments toothed, sometimes turned downward
  along margins
• sandy fields and open woods, sometimes where wet

HABITAT
Dry to moist brushy or grassy meadows, sandy open woods,
mossy places in forests of jack pine and red pine; sometimes
in wet, swampy woods and along shaded streambanks.

DESCRIPTION

FROND 25 cm long or longer, thin and herbacous and not
leathery.

STERILE BLADE Stalked, borne near the ground, broadly triangular in outline; 3-parted, the 3 major divisions stalked,
the terminal portion not usually fully divided; to about 20
cm long and as wide; pinnae margins coarsely toothed,
sometimes reflexed (turned downward).

SPOROPHORE Panicle-like, atop a fleshy stalk; sporangia
yellow, maturing in late summer to fall.

SYNONYMS *Botrychium rugulosum* W.H. Wagner,
Botrychium ternatum auct. non (Thunb.) Sw.

CONSERVATION STATUS *endangered* New York.

NATE MARTINEAU

SIMILAR SPECIES
• **Cut-leaf grape fern** (*Sceptridium dissec-*
  tum) similar but its leaves often turn a
  drab reddish color, and in fall, often turn
  bronze-colored. In ternate grape fern, the
  leaf tends to remain green. Also, the margin teeth of cut-leaf grape fern are serrate
  (sharp and forward-pointing); in *S. rugulo-*
  sum, the teeth are dentate (squarish and
  outward-pointing)
• **Leathery grape fern** (*Sceptridium multi-*
  fidum) similar, but instead of the rounded
  segment lobes of ternate grape fern, the
  sterile blade angular and with margins
  mostly coarsely and irregularly toothed.

NAME 'Rugulosum' refers to tendency of
segments to become more or less wrinkled and convex; ternate refers to the 3-
parted sterile blade.

NOTE Also known as **St. Lawrence grape
fern** due to its presence in St. Lawrence
River valley.

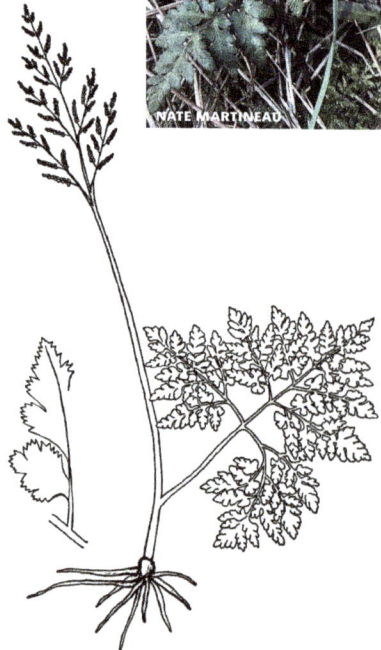

OSMUNDACEAE *Royal fern family*

A SMALL FAMILY OF 6 COSMOPOLITAN GENERA and an estimated 18 species; 2 genera and 3 species in our flora. Members of the family are large clumped ferns of wet to moist places. Fertile and sterile leaves different (*Osmundastrum cinnamomeum*) or the fertile portions different than the sterile portions (*Osmundastrum claytonianum, Osmunda spectabilis*). Rhizomes stout, covered with persistent leaf bases but lacking scales; fronds variously compound; sporangia large, in dense clusters on reduced, stalklike pinnae, these on separate leaves from those bearing sterile pinnae, or on different portions of the leaf.

KEY CHARACTERS
- Large deciduous ferns forming circular clumps.
- Sterile fronds surround inner, more erect fertile fronds.
- Sporangia large, round, in clusters on modified pinnae of fertile fronds.
- Croziers emerge early in spring, covered with woolly white or red-brown hairs.
- Found in swamps, shaded wet depressions, and wet to moist forests.

NOTE Densely matted roots and rhizomes of *Osmunda* or *Osmundastrum* species may form an upright trunk, and are source of the fiber osmundine, used as a growing medium for orchids and other epiphytes.

NAME From Saxon god Osmunder the Waterman, the Saxon equivalent of the Norse god Thor, who supposedly hid his family from danger in a clump of these ferns.

HYBRIDS *Osmunda × ruggii*, uncommon hybrid between interrrupted fern (*Osmundastrum claytonianum*) and royal fern (*Osmunda spectabilis*), reported from Fairfield County, Connecticut.

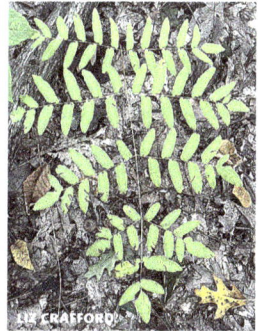

Osmunda spectabilis
ROYAL FERN

KEY TO OSMUNDACEAE | ROYAL FERN FAMILY

1 Fronds of two types, sterile and fertile; sterile fronds broad, 1-pinnate, with woolly tufts at base of pinna-axes; fertile fronds slender, lacking leafy tissue; common, regionwide. . .
. . . . *Osmundastrum cinnamomeum*, CINNAMON FERN, page 198

1 Fronds alike, portions of fronds either sterile or fertile 2

2 Blade 1-pinnate, lacking woolly tufts at base of pinna-axes; fertile pinnae small, in midportion of blade; common, regionwide . . . . *Osmundastrum claytonianum*
. . . . INTERRUPTED FERN, page 200

2 Blade 2-pinnate, with simple oblong pinnules; fertile pinnae at tip of blade; common, regionwide. . . . . *Osmunda spectabilis*, ROYAL FERN, page 196

Osmunda L. ROYAL FERN

Osmunda spectabilis Willd.
ROYAL FERN

DISTRIBUTION CT | MA | ME | NH | NJ | NY | PA | RI | VT

FIELD TIPS
- large, clumped, colony-forming fern of wet places
- sterile fronds similar to compound leaves of locust tree
- fertile pinnae at tips of fronds

HABITAT

Swamps, low woods, marshy meadows, cedar bogs, on hummocks in very wet sites; tolerant of saturated soils.

DESCRIPTION

ROOTSTOCK Erect, large, forming a short hummocky, woody trunk, covered with old roots and winged stipe bases; scales absent.

CROZIER Dark red, covered with brown hairs when very young but soon smooth.

FROND Deciduous, clustered; clustered; sterile and fertile fronds different; 50–180 (–300) cm long.

STIPE Hairy when young, soon smooth, pinkish to green; vascular bundle one, U-shaped, the top of the arms continuing to curl.

RACHIS Grooved, pinkish to green, with reddish to light brown hairs, these soon deciduous.

BLADE With sterile pinnae below and fertile pinnae above, broadly ovate; 2-pinnate.

PINNAE 5 to 9 pairs, not opposite; rotated to the horizontal; on fertile fronds usually the 2 to 4 lowest pairs sterile; pinnules 8–12 pairs plus a terminal pinnule; margins almost entire, the tip of the pinnules finely toothed; veins free, forked.

SORI none; indusium absent; sporangia large, globose, tan or black when mature.

SYNONYM *Osmunda regalis* auct. non L.

NAME In 1753, Linnaeus defined *Osmunda spectabilis* as including both European and Virginian plants; he designated the American plants as a subspecies, but Willdenow made it a species, *O. spectabilis;* in 1810. Gray transferred Willdenow's name to varietal status in 1857. Epithet from Latin for spectacular or showy.

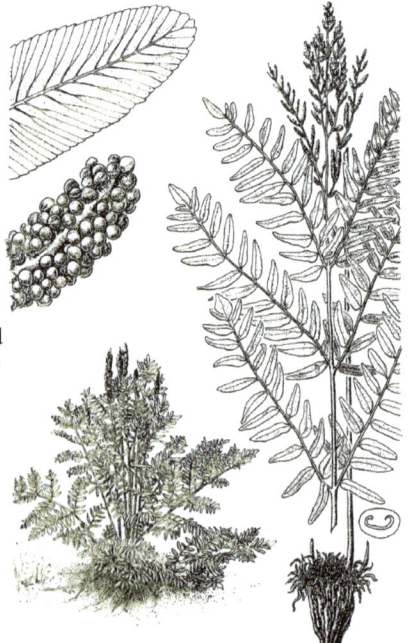

ALEX LOMAS

Osmunda spectabilis ROYAL FERN

Osmundastrum K. Presl **CINNAMON FERN**

Osmundastrum cinnamomeum (L.) C. Presl
CINNAMON FERN

DISTRIBUTION CT | MA | ME | NH | NJ | NY | PA | RI | VT

FIELD TIPS
• large, clumped, colony-forming fern of wet places
• sterile and fertile fronds different; linear fertile fronds surrounded by larger, leafy sterile fronds
• persistent tuft of woolly hairs at base of each pinna

HABITAT
Bogs, acid swamp forests, wet shrub thickets, soils typically sandy or peaty and acidic; usually in partial shade, less common in heavy shade.

DESCRIPTION

ROOTSTOCK Erect, massive, forming a short trunk, occasionally branching, covered with old roots and winged stipe bases; scales absent; roots black, fibrous.

CROZIER Circular, about 2 cm wide, densely covered with whitish hairs which turn cinnamon-brown.

FROND Deciduous, clustered; sterile and fertile fronds different; the outer ring of fronds sterile, arching, longer, to 100 cm long or more; inner fronds fertile, erect, green at first, turning cinnamon brown, developing earlier then withering in early summer and persisting as a hairy dried stalk.

STIPE With rusty woolly hairs when young, soon smooth; vascular bundle 1, U-shaped, the top of the arms continuing to curl inward.

RACHIS Green; sparsely covered with cinnamon-colored wool when young, becoming smooth apart from a tuft of brownish hairs at base of each pinna.

STERILE BLADE Elliptic to oblong, arching, 1-pinnate-pinnatifid, pinnae broadly oblong.

FERTILE BLADE Pinnae small, angled upward, bearing sporangia.

PINNAE 20 to 25 pairs, rotated to the horizontal; pinnules obtuse; costae and rachis shallowly grooved above; margins entire; veins free, forked 1–3 times; rusty wool on pinnae and rachis soon deciduous; tuft of tan hairs on underside of pinna base near rachis tends to persist.

SORI None; indusium absent; sporangia large, globose, greenish when young, tan or black when mature.

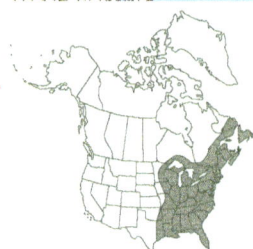

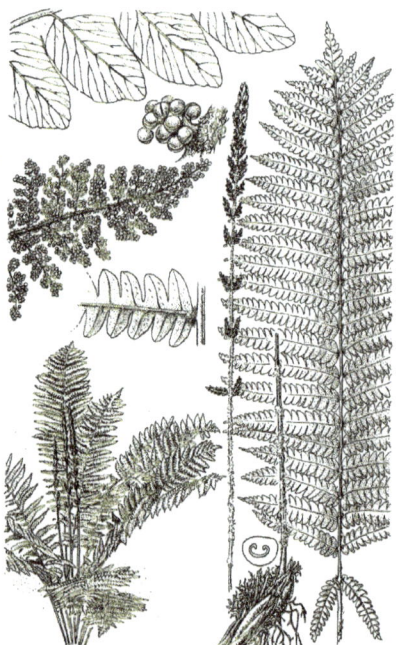

SYNONYM *Osmunda cinnamomea* L.

SIMILAR SPECIES **Interrupted fern** (*Osmundastrum claytonianum*), **ostrich fern** (*Matteuccia sruthiopteris*) and **Virginia chain fern** (*Anchistea virginica*) are similar ferns of wet habitats, but all lack the tan woolly hairs at pinnae base.

NOTE In early spring, young uncoiling sterile fronds ('fiddleheads'), were once considered edible but now known to be carcinogenic (the best edible fiddleheads are those of ostrich fern, page 160). The crisp, tender, central part of the crown, with its slightly nutty flavor, were also sometimes eaten raw.

NAME Specimens of this fern were sent to Linnaeus from Maryland and named by him in 1753.

BOTANYGIRL

CHARLES DE MILLE-ISLES
Osmundastrum cinnamomeum CINNAMON FERN

Osmundastrum claytonianum (L.) Tagawa
INTERRUPTED FERN

DISTRIBUTION CT | MA | ME | NH | NJ | NY | PA | RI | VT

FIELD TIPS
• large deciduous fern
• fronds with small fertile pinnae near middle of blade
• sterile fronds smaller, arching
• no tuft of hairs at base of pinnae as in cinnamon fern
• moist to wet forests

HABITAT

Low places in mesic to wet forests, on hummocks in swamps, roadsides, in shade to partial shade, less tolerant of wet conditions than cinnamon fern or royal fern.

DESCRIPTION

ROOTSTOCK Erect and also creeping and branching, large, covered with old roots and winged stipe bases; scales absent; roots black, fibrous.

CROZIER Circular, about 2 cm wide, densely covered with whitish hairs when young, turning brownish; very similar to those of *O. cinnamomeum*.

FROND Deciduous, clustered; sterile and fertile fronds different; outer ring of fronds sterile, arching, 40–180 cm long and longer than the fertile fronds; inner fronds fertile, erect, with widely spaced pinnae.

STIPE Hairy when young, soon smooth; vascular bundle one, U-shaped, the top of the arms continuing to curl.

RACHIS Green; sparsely covered with reddish to light brown woolly hairs, at least when young, but no tufts of hairs at base of pinna as in cinnamon fern (*O. cinnamomeum*).

STERILE BLADE Elliptic to oblong, arching, 1-pinnate-pinnatifid.

FERTILE BLADE With 2–7 middle pinnae pairs reduced in size, these bearing the sporangia, and soon withering.

PINNAE Broadly oblong, 20 to 30 pairs, rotated to the horizontal; margins entire, tip rounded; veins free, once-forked.

SORI None; indusium absent; sporangia large, globose, greenish young, tan or black when mature.

SYNONYMS *Claytosmunda claytoniana* (L.) Metzgar & Rouhan, *Osmunda claytoniana* L.

NAME John Clayton collected this fern in Virginia in the early 1700s, and it was named in his honor by Linnaeus in 1753.

JAY BRASHER

AARON GUNNAR

Osmundastrum claytonianum **INTERRUPTED FERN**

POLYPODIACEAE *Polypody family*

A LARGE FAMILY OF 65 GENERA and an estimated 1652 mostly tropical species; one genus and two species in our region. Most ferns in the family grow on trees or rock rather than in soil.

KEY CHARACTERS
- Evergreen, colony-forming ferns of rock cliffs, boulders.
- Leaf blades pinnatifid.
- Sterile and fertile fronds alike.
- Sori round, indusia absent.

NAME From Greek, *polys,* many, and *pous,* foot, perhaps referring to the numerous stipe bases, or to the many
   knoblike branches of the rhizome.

Polypodium L. POLYPODY

EVERGREEN, COLONY-FORMING FERNS of rock outcrops; rhizome creeping, scaly; fronds pinnatifid, the lobes rounded at tip, stipes scaly, with 3 vascular bundles; sori round, indusium absent. Genus includes ca. 160 species worldwide; 2 species in our flora.

KEY CHARACTERS
- Evergreen, colony-forming fern of rock outcrops.
- Leaf blades pinnatifid, leathery, without scales on underside.
- Sterile and fertile fronds alike.
- Sori round, indusia absent.
- Fronds wither in drought, quickly becoming green following rain.

NAME From Greek, *polys,* many, and *pous,* foot, perhaps referring to the numerous stipe bases, or to the many
   knoblike branches of the rhizome.

KEY TO *POLYPODIUM* | POLYPODY

1   Leaf blade widest just above base, lobes narrowed at tip (especially lower lobes); rhizome scales a uniform golden brown; regionwide . . . . *Polypodium appalachianum*
. . . . APPALACHIAN POLYPODY, page 203
1   Leaf blade widest near middle, the blades more evenly wide; lobes blunt-tipped; rhizome scales of 2 shades of brown; common, regionwide . . . . *Polypodium virginianum*
. . . . ROCK POLYPODY, page 205

CONNOR SITES-BOWEN   *Polypodium appalachianum* APPALACHIAN POLYPODY

Polypodium appalachianum Haufler & Windham
APPALACHIAN POLYPODY

DISTRIBUTION CT | MA | ME | NH | NJ | NY | PA | RI | VT

FIELD TIPS
- evergreen, colony-forming fern of rock outcrops and rocky slopes
- blade pinnatifid, broadest at base, leathery
- sori round, indusia absent

HABITAT

On rock or thin soil over rock; cliffs, boulders, talus.

DESCRIPTION

ROOTSTOCK Long-creeping, branching, with whitish covering; scales lanceolate, mostly uniformly golden brown.

FROND Evergreen, to 35 cm long, sterile and fertile fronds alike.

STIPE Jointed at base, straw-colored, smooth or with scattered light-brown scales; vascular bundles 3.

RACHIS Straw-colored, smooth; sunken below plane of upper surface, somewhat raised on undersurface.

BLADE Pinnatifid, oblong to narrowly lanceolate, usually widest near base, leathery; rachis sparsely scaly on underside, upperside smooth.

PINNAE Linear, entire to slightly toothed, broadest near their base; veins free.

SORI Round, midway between margin and midrib to nearly marginal; on all but lowest pinnae of fertile fronds; indusium absent; sporangia yellow, then brown when mature.

SYNONYM *Polypodium virginianum* forma *acuminatum* (Gilbert) Fern.

SIMILAR SPECIES Distinguished from **rock polypody** (*Polypodium virginianum*) by being widest near base of blade and narrower at tip, the pinnae tips tend to be more pointed than in *P. virginianum* (at least near base of blade), and rhizome scales uniformly golden brown.

NAME Also called **rock cap fern** for its habit of densely covering rocks and boulders.

ALINA MARTIN

MATT BERGER

Polypodium appalachianum APPALACHIAN POLYPODY

Polypodium virginianum L.
ROCK POLYPODY

DISTRIBUTION CT | MA | ME | NH | NJ | NY | PA | RI | VT
FIELD TIPS
• evergreen, colony-forming fern of rock outcrops
• blade pinnatifid, leathery
• sori round, indusia absent

HABITAT
Sandstone and calcareous bluffs, rock outcrops, ledges, boulders and mossy talus, rocky soil on steep slopes, and on decaying stumps and logs; usually in partial shade.

DESCRIPTION
ROOTSTOCK Long-creeping at ground surface, ropelike and spongy, occasionally branching, often partly exposed and mat-like, producing rows of fronds; densely covered with brown scales.

FROND Evergreen, last year's fronds withering as this year's appear in early summer, sterile and fertile fronds alike; mostly 10–40 cm long.

STIPE Jointed at base, straw-colored, smooth or with scattered light-brown scales; vascular bundles 3.

RACHIS Straw-colored, smooth; sunken below plane of upper surface, somewhat raised on undersurface.

BLADE Pinnatifid, cut almost to rachis, oblong to narrowly lance-shaped, usually widest near middle, occasionally at or near base, deep green and leathery, upper surface lustrous.

PINNAE 9 to 18 pairs, linear, tip rounded; margins entire to round-toothed; veins free, forked, obscure.

SORI Round, midway between margin and midrib to nearly marginal, at vein ends; on all but lowest pinnae of fertile fronds; indusium absent; sporangia yellow to brown at maturity; paraphyses (branching structures among the sporangia) present.

SYNONYMS *Polypodium vulgare* auct. non L. p.p., *Polypodium vulgare* var. *virginianum* (L.) Eaton

NOTES Henry David Thoreau referred to the "fresh and cheerful communities" of polypody in early spring. The rhizome has a strong licorice taste.

NAME In 1753, Linnaeus recognized a European *Polypodium vulgare* and an American *P. virginianum* (*virginianum* describes the area of Virginia).

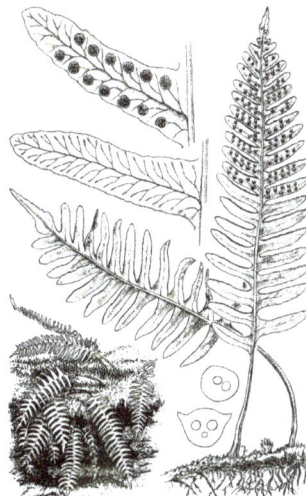

BOB SCHWARTZ

Polypodium virginianum ROCK POLYPODY

PTERIDACEAE *Maidenhair fern family*

SMALL TO MEDIUM FERNS, mostly growing on rock. Fronds compound, sori linear along or near pinnae margin, covered by false indusium formed by inrolled margin. Worldwide, the family includes 53 genera and an estimated 1211 species, most commonly in tropical and arid regions; 6 genera in our flora.

KEY CHARACTERS

- Typically growing on rocks (except for *Adiantum pedatum* which also grows in soil).
- Rhizomes scaly.
- Sori linear along pinnae margin, covered by false indusium formed by inrolled margin.

SUBFAMILIES

This large family sometimes subdivided into 4 subfamilies, all of which occur in the northeast:

- CHEILANTHOIDEAE – *Myriopteris*
- CRYPTOGRAMMOIDEAE - *Cryptogramma*
- PTERIDOIDEAE - *Pellaea, Pteris*
- VITTARIOIDEAE - *Adiantum, Vittaria*

NAME From Greek, *pteris,* fern, derived from *pteron,* wing or feather, for the closely spaced pinnae, the leaves resembling feathers.

ADDITIONAL NORTHEASTERN SPECIES

- **Spider brake** (*Pteris multifida* Poir.), introduced from Asia in southeastern USA (where usually found in cracks of rock walls); in the northeastern region reported for New York. Rootstock short-creeping, densely scaly, scales reddish brown. Frond deciduous in our region, the sterile fronds leafier than the fertile. Stipe dark brown to straw-colored at base, sparsely scaly at base, smooth above; vascular bundle 1. Leaf blade 1-pinnate, bright green, smooth; pinnae 3 to 7 pairs, linear, lower pinnae 1-pinnate, upper not divided, joined to the winged rachis; fertile pinnae contracted; veins conspicuous, free, simple or once-forked. Sori continuous, near margin; indusium false, formed by reflexed margin of pinnae. Synonym: *Pycnodoria multifida* (Poiret) Small.

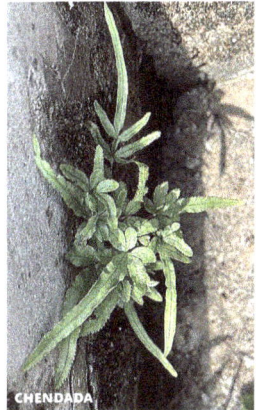

Pteris multifida
SPIDER BRAKE

KEY TO PTERIDACEAE | MAIDENHAIR FERN FAMILY

1 Plant a free-living gametophyte, resembling a thallose liverwort; rare in dark, moist recesses in rock overhangs; rare in NY, PA . . . . *Vittaria appalachiana*
. . . . SHOESTRING FERN, page 220

1 Plant a sporophyte, consisting of a stem, rhizome, corm, or crown producing well-developed leaves. 2

2 Plants of moist forests; sori distinct, short, mostly not merged together; regionwide . . . .
. . . . *Adiantum*, MAIDENHAIR FERN, page 208

2 Plants of rock cliffs and boulders; sori usually merged as a marginal band 3

3 Segments of frond hairy, sometimes bead-like; CT, NJ, PA . . . . *Myriopteris lanosa*
. . . . HAIRY LIP FERN, page 214

3 Segments of frond without hairs, not bead-like. **4**

4 Petioles green to straw-colored for at least the upper 1/3, rachis green; sterile and fertile fronds different, the fertile longer than the sterile and with narrower segments; CT, MA, ME, NH, NY, PA, VT. . . . . *Cryptogramma stelleri*, FRAGILE ROCKBRAKE, page 212

4 Petioles and rachis dark brown to almost black throughout; leaves uniform; regionwide but absent from ME . . . . *Pellaea*, CLIFFBRAKE, page 216

Adiantum L. MAIDENHAIR FERN

ADIANTUM INCLUDES ABOUT 150 SPECIES WORLDWIDE, mostly of tropical regions. North America is home to 9 species; 3 species in our flora, one of which (*Adiantum aleuticum*) is disjunct in the northeast region from western North America (see below). Maidenhair ferns are deciduous, monomorphic ferns with slender purplish-black, glossy stipes and rachises; pinnules somewhat resemble the leaves of the *Ginkgo biloba* tree. Maidenhair ferns are terrestrial and typically found in shaded, rocky woods, often along streambanks and seepages.

KEY CHARACTERS
• Delicate fern of moist woods, blade fan-shaped to round in outline.
• Pinnules thin, midrib absent; main vein located along lower pinnule margin.
• Stipe and rachis glossy chestnut brown to purple-black, brittle.
• Sori along margins, covered by and attached to the inrolled edge of pinnule.

NAME From Greek, *a*, not, and *diaine*, to wet, referring to the water-shedding properties of the leaves. 'Maidenhair' may refer to either the slender black stipes or to the fine, black, fibrous roots.

KEY TO *ADIANTUM* | MAIDENHAIR

1 Ultimate segments of fronds ± oblong, rounded to obtuse at the tip, in the same plane with the axis of the blade, herbaceous, bright green to yellow-green; leaf blades lax and arching; fern usually in rich, mesic forests; common, regionwide . . . . *Adiantum pedatum* . . . . NORTHERN MAIDENHAIR FERN, page 210

1 Ultimate segments of fronds ± triangular, acute to obtuse at tip, often twisted out of the plane of the axis of the blade, firm-herbaceous to paper-like, medium green to blue-green (but may be bright green when growing in shade); leaf blades usually stiff and upright (or somewhat lax in shaded plants); ferns of serpentine rocks and soil; uncommon in ME and VT. **2**

2 Medial ultimate segments of fronds on stalks mostly less than 1 mm long; false indusia 1–3 mm long; rhizomes often crowded and branched, with internodes 1–2 mm long; uncommon in ME and VT . . . . *Adiantum aleuticum* . . . . WESTERN MAIDENHAIR FERN, page 209

2 Medial ultimate segments of fronds on stalks 0.5–1.5 mm long (most stalks longer than 1 mm); false indusia 2–5 mm or more long; rhizomes only occasionally branching (and rarely crowded), with internodes 4–7 mm long; uncommon in ME and northern VT. . . . . *Adiantum viridimontanum*, GREEN MOUNTAIN MAIDENHAIR FERN, page 209

ADDITIONAL NORTHEASTERN SPECIES

- **Western maidenhair fern** [***Adiantum aleuticum*** (Rupr.) Paris], a fern mostly of western North America, is reported from northern Vermont and western Maine. Very similar to the much more common **northern maidenhair** (*A. pedatum*), and sometimes treated as a variety of that species. However, in *A. aleuticum,* pinnae sometimes oriented vertically; pinnae in *A. pedatum* always horizontal. Other differences include a middle pinna larger than the other pinnae; green, rather than red new growth; and presence of a single fan-shaped pinnule on the rachis between the fork and the first pinnae; this pinnule is rarely present on *A. pedatum.*

- **Green Mountain maidenhair fern** (*Adiantum viridimontanum* Paris); serpentine cliffs and talus, rocky woods and forest edges on serpentine bedrock; endemic to Maine, northern Vermont, and southern Quebec; see key for distinguishing features.

CRICKET RASPET

Adiantum aleuticum
WESTERN MAIDENHAIR

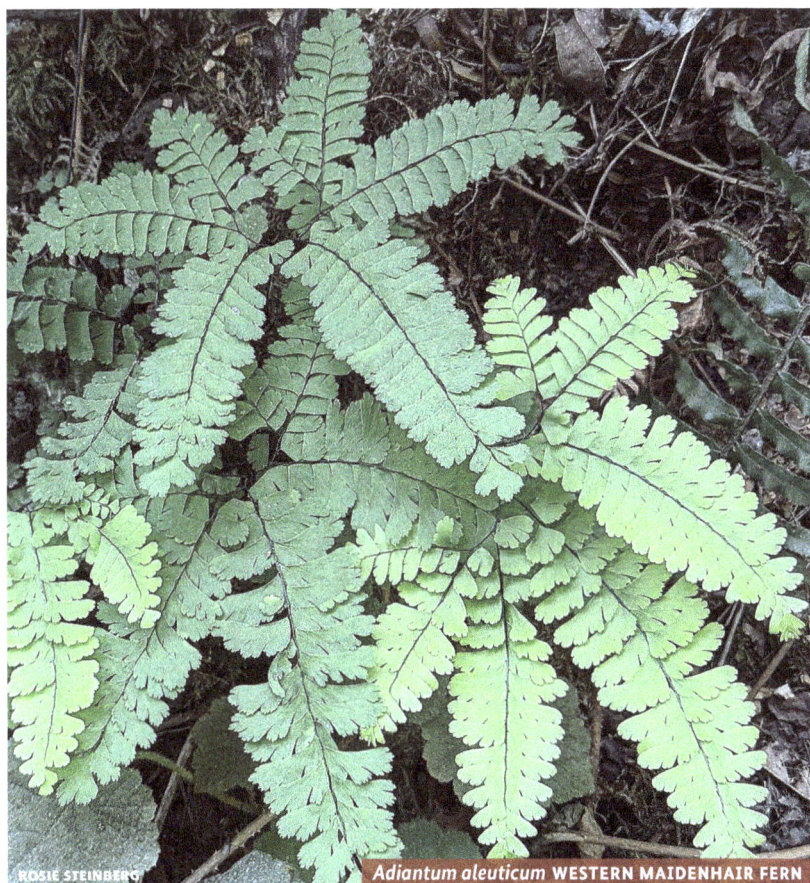

ROSIE STEINBERG

Adiantum aleuticum **WESTERN MAIDENHAIR FERN**

Adiantum pedatum L.
NORTHERN MAIDENHAIR FERN

DISTRIBUTION CT | MA | ME | NH | NJ | NY | PA | RI | VT

FIELD TIPS
- delicate deciduous fern of moist woods, forming small colonies
- unique fan-shaped blade
- false indusium formed by inrolled pinnule margin
- stipe and rachis purplish black

HABITAT

Rich deciduous forests, often near seeps or streambanks; rocky ravines.

DESCRIPTION

RHIZOME Short-creeping and branching, woody; scales bronze-colored.

CROZIERS Clustered, delicate, wine-red.

FROND Deciduous, separate but close together along the rhizome; sterile and fertile fronds alike; ca. 40 cm long by 30 cm wide.

STIPE Purplish black, slender, grooved above, smooth, occasionally waxy or with or a few scales at base; vascular bundle single, U- or V-shaped.

RACHIS Purple-brown to black, smooth.

BLADE Fan-shaped or nearly circular from the branched rachis, the primary divisions then pinnately divided; held horizontally, at 90° to the stipe; smooth.

PINNAE 6 to 10 pairs, oblong, those closest to the stipe longer, becoming smaller outwards; pinnules alternate, incised on upper margin; margins crenate; veins free, several times forked from main vein along lower margin, the midrib absent.

SORI Elongate, near upper margin of pinnule; indusium false, formed by reflexed margin of pinnule; sporangia yellow or yellowish brown.

NOTE The dark, wiry stipes were used to accent larger tan-colored reeds in basket-making by Native Americans. In earlier times, maidenhairs had many medicinal uses.

SIMILAR SPECIES Southern maidenhair fern (*Adiantum capillus-veneris*), a fern of the southern USA, is not known from our region, but extends northward to Kentucky and Missouri; unlike *A. pedatum*, the stipe is not forked.

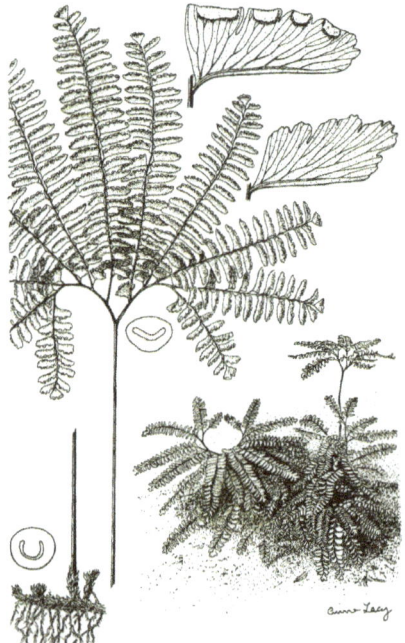

Sent to Linnaeus from Canada and Virginia, and named by him in 1753. From Latin, *pedatus,* having feet, referring to the branching of the blade.

ERIC BARATTA

Adiantum pedatum NORTHERN MAIDENHAIR FERN

Cryptogramma R. Br. ROCKBRAKE

SMALL EVERGREEN OR DECIDUOUS, DIMORPHIC FERNS found on rock; fertile fronds larger than sterile. Genus includes about 11 species, mostly at higher elevations in North America, Europe, the Andes, and parts of Asia; four species in North America, one species in our flora.

KEY CHARACTERS
- Small evergreen or deciduous ferns of rock cliffs.
- Fronds dimorphic, fertile fronds taller than sterile fronds.
- Leaf blades 2-pinnate.
- Stipe dark brown at base, light brown to green upwards, grooved on upperside.
- Sori near margins of fertile pinnules, the margins inrolled and forming a false indusium.

NAME From Greek, *kryptos,* hidden, and *gramme,* line; referring to hidden lines of sori under the reflexed leaf margin. 'Rockbrake' refers to the typical rocky habitat, brake is used as a general term for fern. Also known as 'parsley fern' as sterile fronds of some species resemble parsley.

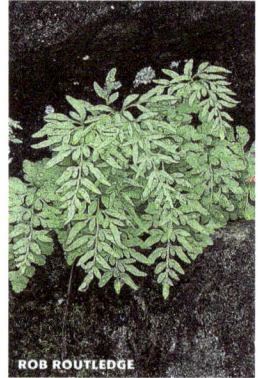
ROB ROUTLEDGE

Cryptogramma stelleri (Gmel.) Prantl
FRAGILE ROCKBRAKE

DISTRIBUTION CT | MA | ME | NH | NJ *historical* | NY | PA | VT
FIELD TIPS
- small fern of rock cliffs and ledges
- sterile and fertile fronds deciduous, much different in form
- margins inrolled over sori to form false indusium

HABITAT
Moist, shaded, rocks and cliffs; wet or seepy crevices and ledges of limestone, calcareous shale and sandstone.

DESCRIPTION
ROOTSTOCK Long-creeping, slender (to 1 mm wide), seldom branching, hairy, with a few transparent scales.

FROND Deciduous, dying by late summer; sterile and fertile fronds unalike, the sterile fronds shorter than the fertile; to about 18 cm long.

STIPE Brown at base, green upwards, smooth or with a few hairs at base; vascular bundles 2.

RACHIS Green, smooth.

BLADE Fertile fronds lance-shaped, 2-pinnate, the sterile less divided, pinnate-pin-

natifid; bright green, thin-textured and nearly translucent, smooth.

PINNAE 5–6 pairs, the fertile pinnae smaller; margins entire, inrolled on fertile segments; veins free, forked.

SORI Elongate, near margins; indusium false, formed by the strongly inrolled margin; sporangia yellow.

CONSERVATION STATUS *endangered* Connecticut, Massachusetts New Hampshire, Pennsylvania; *threatened* Maine, New Jersey (historical record only).

SIMILAR SPECIES Somewhat like a small cliffbrake (Pellaea), but unlike that genus, the sterile and fertile fronds very different, and rachis greenish rather than dark brown.

NAME First named *Pteris stelleri* by Gmelin based on plants from Siberia in 1768. Michaux named identical Canadian plants *Pteris gracilis* in 1803. In 1823, R. Brown proposed genus *Cryptogramma,* and Prantl placed the species in that genus in 1882.

Species named in honor of German naturalist George Wilhelm Steller (1709–1746), who accompanied the Bering expedition of 1741 and who is considered the first European to set foot in what is now Alaska; his name is used with other species, including Steller's jay and Steller's sea lion.

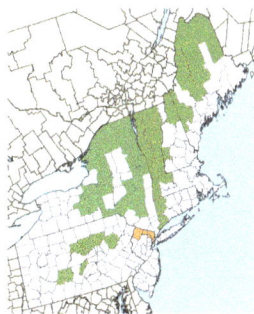

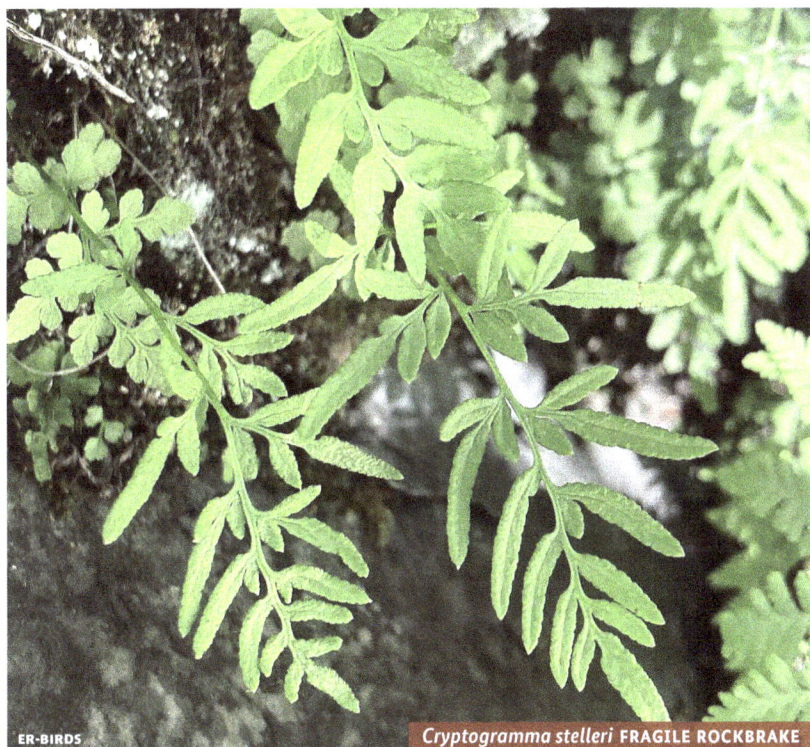

Cryptogramma stelleri **FRAGILE ROCKBRAKE**

ER-BIRDS

Myriopteris Fée LIP FERN

SMALL, EVERGREEN (OR NEARLY SO), MONOMORPHIC FERNS, usually woolly hairy, at least on underside of blade. Plants well-adapted to dry conditions of their typical rock cliff habitats, and during drought, may curl up and appear lifeless, then quickly greening after rain. *Myriopteris* includes ca. 45 species, mostly in the Western Hemisphere; a few species occur in Europe, Asia, Africa, Australia and the Pacific Islands. There are 28 species in North America, mostly in deserts of the southwestern United States and northern Mexico; 1 species in our flora.

KEY CHARACTERS
• Small, nearly evergreen ferns of rock outcrops.
• Stipe brown to black.
• Leaf blade mostly 2-pinnate-pinnatifid, leathery, finely hairy and/or scaly below, finely hairy or mostly smooth above.
• Pinnule lobes sometimes rounded and beadlike.
• Sporangia at ends of veins in sori-like clusters, more or less continuous along margins of pinnules, covered by false indusium formed by inrolled margin.

NOTE Separated from other small cliff-dwelling ferns by the small, hairy beadlike pinnule lobes and dark brown or black stipe and rachis. Unlike most ferns, many *Myriopteris* have noncircinate vernation, that is, the emerging bud is not at the center of a coil, in contrast to the more usual unfurling form of a crozier or fiddlehead form.

NAME From Greek, *cheilos,* lip, and *anthos,* flower; a lip-like false indusium covers the sporangia.

Myriopteris lanosa (Michx.) Grusz & Windham
HAIRY LIP FERN

DISTRIBUTION CT | NJ | NY *historical* | PA
FIELD TIPS
• small loosely clumped fern in shallow soil on sandstone outcrops
• underside of blade with woolly tan-gray hairs
• stipe brown, hairy
• plants dry up in drought

HABITAT
Dry, exposed, sandstone bluffs and ledges.

DESCRIPTION
ROOTSTOCK Short-creeping, branched, without hairs; scales brown, narrowly lance-shaped.
FROND Nearly evergreen, becoming dry and dormant in droughts, sterile and fertile fronds alike; to 40 cm long by 5 cm wide.

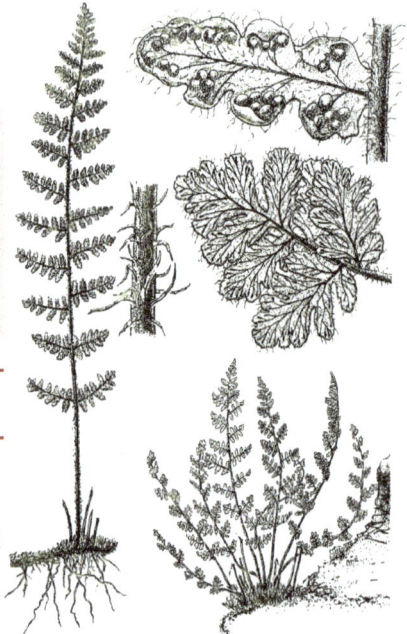

STIPE Chestnut brown, the color extending into the rachis and costae, with many small white-grayish hairs; scales absent; vascular bundles 1.

RACHIS Covered in white hairs, turning grayish when mature.

BLADE Lance-shaped, 2-pinnate-pinnatifid at base, less cut above; underside woolly with jointed tan-gray hairs, upper surface less hairy.

PINNAE Gray-green; 12 to 14 pairs; pairs opposite and widely separated at base, closer together and sub-opposite to alternate above; pinnules opposite, pinnate at bottom of blade, pinnatifid above; underside sparsely reddish hairy; veins free, forked.

SORI Linear, discontinuous, near margin; indusium false, formed by the slightly inrolled margin; sporangia black.

SYNONYMS *Cheilanthes lanosa* (Michx.) D.C. Eaton, *Cheilanthes vestita* (Spreng.) Sw.

CONSERVATION STATUS *endangered* Connecticut, New York (historical records only).

NAME Discovered by Michaux in the mountains of Tennessee and North Carolina and named *Nephrodium lanosum* in 1803; assigned to genus *Cheilanthes* by Watt in 1874. Epithet from Latin, *lana,* woolly or soft-hairy.

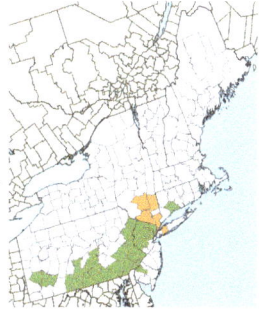

Myriopteris lanosa **HAIRY LIP FERN**

Pellaea Link **CLIFFBRAKE**

CLIFFBRAKES ARE SMALL EVERGREEN FERNS of rocky, often calcareous habitats. Sterile and fertile fronds similar (*Pellaea glabella*) or somewhat different (*P. atropurpurea*). Sori near pinna margin and covered by inrolled edge forming a false indusium. *Pellaea* includes ca. 40 species, mostly in the Western Hemisphere; 15 species in North America; 2 in our flora. In the United States, most species of *Pellaea* are found in arid regions of the Southwest.

KEY CHARACTERS
• Small evergreen ferns of rock cliffs and talus.
• Rhizomes short-creeping, scaly.
• Sterile and fertile fronds similar or slightly different.
• Leaf blade gray-green or blue-green, 2-pinnate at base, less so above; tipped with a terminal leaflet.
• Stipe shiny, dark brown or purple.
• Sori marginal, covered by false indusium formed by edge of inrolled pinna.

NAME From Greek, *pellos,* dusky, referring to the dull, bluish-gray pinnae. 'Cliffbrake' refers to the typical rock crevice habitat, 'brake' is a general term for fern.

KEY TO *PELLAEA* | CLIFFBRAKE

1   Stipe and rachis without short, incurved hairs; lower pinnule surface smooth or with a few hairlike scales; CT, NJ, NY, PA, VT ...... *Pellaea glabella*, SMOOTH CLIFFBRAKE, page 218

1   Stipe and rachis with short, incurved hairs on upper surfaces; lower pinnule surface with scattered hairlike scales; regionwide, but absent from ME .... *Pellaea atropurpurea* .... PURPLE-STEM CLIFFBRAKE, page 216

Pellaea atropurpurea (L.) Link
PURPLE-STEM CLIFFBRAKE

DISTRIBUTION CT | MA | NH | NJ | NY | PA | RI | VT
FIELD TIPS
• small evergreen fern of rock and talus
• sterile and fertile fronds not alike, leathery, dull gray-green
• stipe and rachis dark purple-brown to nearly black, hairy
• sori covered by inrolled pinnule margin

HABITAT
Dry, exposed limestone outcrops, ledges and talus slopes; also found on shale and sandstone; less common on exposed vertical cliffs (that habitat more typical of *Pellaea glabella*).

DESCRIPTION
ROOTSTOCK Short-creeping, branched; densely covered with brown scales.

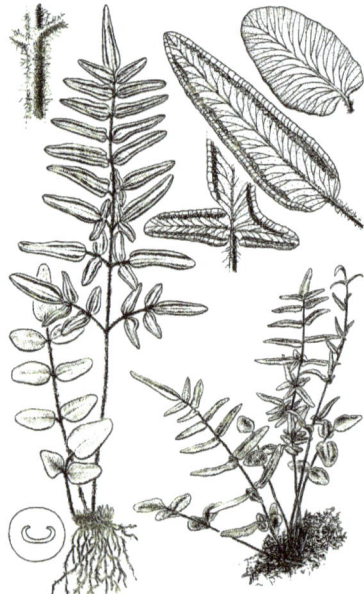

FROND Evergreen, somewhat dimorphic, sterile shorter and less divided than fertile fronds; 50 cm long by 18 cm wide.

STIPE Dark purple to black, shiny, with incurved hairs; base with scales, these uniformly brown; vascular bundle 1.

RACHIS Dark purplish brown; usually covered with whitish hairs.

BLADE Long triangular in outline, leathery, gray-green; usually 2-pinnate at base, but sometimes more or less.

PINNAE 5 to 9 pairs, lower pinnae stalked, upper sessile, with a terminal pinna like the upper lateral ones; pinnules sessile or nearly so, sometimes with 1 or 2 basal lobes; veins obscure.

SORI Elongate or linearly joined, submarginal; indusium false, margins inrolled; sporangia brown.

CONSERVATION STATUS *endangered* New Hampshire, Rhode Island.

SIMILAR SPECIES Similar to **smooth cliffbrake** (*Pellaea glabella*), but differs in having hairs on rachis and costae; fronds also taller, more upright and more divided. Distinguished from **fragile rockbrake** (*Cryptogramma stelleri*) by its leathery pinnules and hairy stipe and rachis.

NAME Discovered by John Clayton in Virginia and described by Gronovius in about 1743; named by Linnaeus *Pteris atropurpurea* in 1753. When Link founded genus *Pellaea* in 1841, he included this species. From Latin, *atropurpureus*, dark purple, referring to color of stipe and rachis.

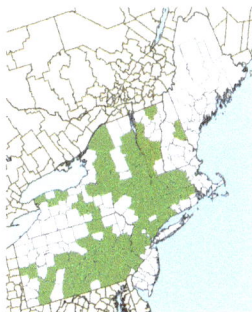

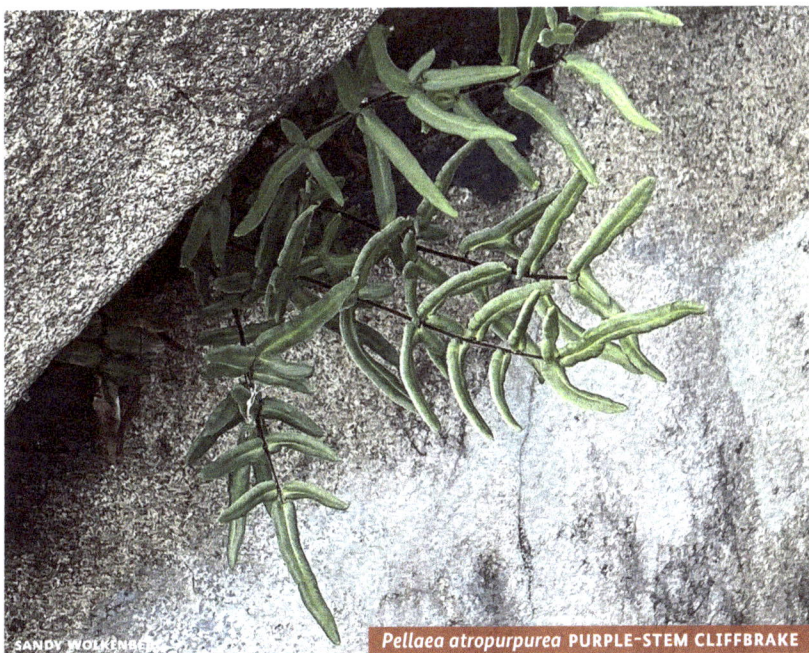

Pellaea atropurpurea **PURPLE-STEM CLIFFBRAKE**

Pellaea glabella Mett. ex Kuhn
SMOOTH CLIFFBRAKE

DISTRIBUTION CT | NJ | NY | PA | VT

FIELD TIPS
- small evergreen fern of rock cliffs
- sterile and fertile fronds similar, leathery, pale bluish green
- stipe dark reddish brown, shiny, without hairs
- sori covered by inrolled pinnule margin

HABITAT
Ledges and crevices of cliffs of limestone, dolomite and sandstone; in sun to light shade.

DESCRIPTION

ROOTSTOCK short-creeping or erect, branched; scales orange-brown.

CROZIER Sparsely hairy at emergence, then smooth.

FROND Evergreen, clustered, sterile and fertile fronds similar but sterile fronds typically shorter and less divided; to 35 cm long by 8 cm wide.

STIPE Dark reddish brown, shiny, smooth or with a few hairs, vascular bundle 1.

RACHIS Brown, smooth or very sparsely hairy.

BLADE Usually 2-pinnate at the base, less above, leathery, pale bluish green,

PINNAE 4 to 9 pairs, lower pinnae stalked, upper sessile, with a terminal pinna like the upper lateral ones; pinnules sessile or nearly so, sometimes with 1 or 2 basal lobes; margins entire; veins obscure, forked.

SORI Elongate or linearly joined, submarginal; indusium false, margins inrolled to cover sori; sporangia pale brown.

CONSERVATION STATUS *endangered* Connecticut, *threatened* New York.

SYNONYM *Pellaea atropurpurea* (L.) Link var. *bushii*

SIMILAR SPECIES Similar to **purple-stem cliffbrake** (*Pellaea atropurpurea*), but smaller, the sterile and fertile fronds alike, and stipe and rachis without hairs.

NAME For many years this species mistaken for depauperate specimens of *Pellaea atropurpurea*. Plants sent from Missouri to Mettenius were recognized by him to be distinct, and his suggested name, *Pellaea glabella*, was published by Kuhn in 1869. Latin, *glabella*, hairless or smooth.

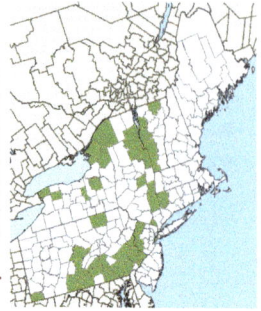

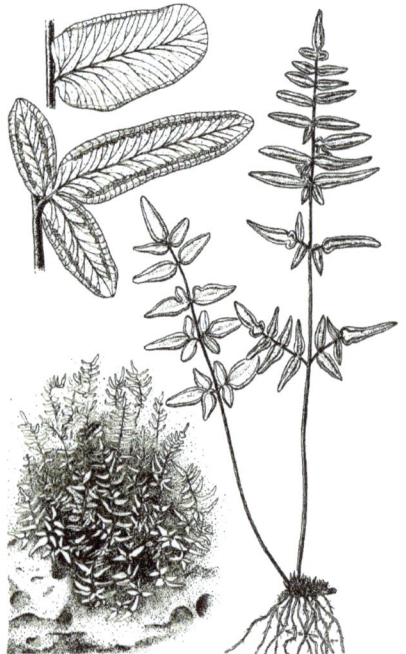

MATT BERGER

Pellaea glabella SMOOTH CLIFFBRAKE

Vittaria Sm. SHOESTRING FERN

Vittaria appalachiana Farrar & Mickel
APPALACHIAN SHOESTRING FERN

DISTRIBUTION NY | PA

FIELD TIPS
• restricted to dark, moist rock grottoes and small caves
• resembles a thallose liverwort
• may form dense colonies
• only found in gametophyte form

HABITAT
On ceilings, walls and floors of cool, perennially moist, heavily shaded or dark cavities, rock shelters or overhangs in non-calcareous rock (sandstone or conglomerate); sometimes requiring a flashlight to see.

DESCRIPTION
Unique among our ferns in that the sporophyte generation (the typically fern-like stage of the life cycle) is essentially absent, and plants occur as asexually-reproducing gametophytes (as also the case in *Crepidomanes intricatum,* page 152). Small (less than 1 cm long), abortive sporophytes have been recorded from Ohio, and have also been produced in culture.

Plants small, 1–3 cm long, with branches at tips bearing small gemmae (several-celled plantlets); gemmae shed to form new plants, and colonies up to several square meters may form in suitable habitat.

CONSERVATION STATUS *endangered* New York, *threatened* Pennsylvania.

NOTE Formerly included within Vittariaceae (Shoestring Fern Family), but that family no longer considered valid. Worldwide, 7 species; 3 species in continental USA. *V. lineata* occurs in both sporophyte and gametophyte forms, and occurs in Florida, southern Alabama, Georgia, and South Carolina. It is a common epiphyte on palm trees in Florida.

NAME Genus named by James Edward Smith in 1793. *Vittaria* from Latin, *vitta,* a band or ribbon. Our plants first recognized as the gametophyte of a fern by A.J. Sharp in 1930, and assigned to genus *Vittaria* by Wagner and Sharp in 1963. Farrar and Mickel described the species in 1991. The common name refers to the hanging, long-linear form of other species of *Vittaria.*

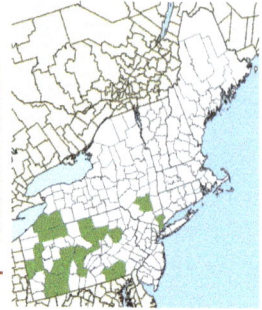

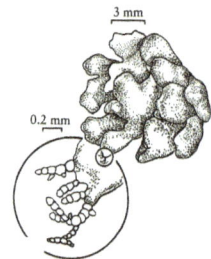

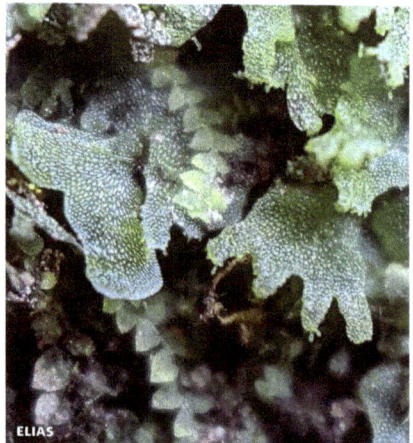

SALVINIACEAE *Water fern family*

A SMALL FAMILY OF AQUATIC FERNS, with two genera and an estimated 21 species.

ADDITIONAL SPECIES **Water Spangles (*Salvinia minima***
Baker), an introduced aquatic weed reported in our region
from Suffolk County, MA, and Rockland County, NY; well-
established in Florida, where considered invasive, and in
other southeastern states. This small fern, native to South
America and introduced to the United States in the 1920s–
30s, can form thick mats across the water surface to the
detriment of other organisms. Leaves of *Salvinia minima*
are small and oval, about 0.5–2 cm in length; the leaves
grow in joined sets of three, with two leaves floating on
the surface and one leaf dissected, and hanging underneath.

Salvinia minima
WATER SPANGLES

Azolla Lam. **MOSQUITO FERN**

SMALL, FREE-FLOATING AQUATIC FERNS of quiet or stagnant water of ponds, lakes,
marshes and streams, may form large matlike colonies covering the water surface. Two native species are recognized in the USA: *Azolla cristata* Kaulfuss (eastern and midwestern plants) and *Azolla filiculoides* Lam. (western USA from
Washington to west Texas).

KEY CHARACTERS
- One of the world's smallest ferns but most important economically.
- Mat-forming, floating aquatic fern of quiet water.
- Plants small, reddish when in full sun to bright green in shade, the segments easily
  broken or dispersed to form new colonies.
- Pinnae small, scalelike.
- Spores of 2 types (but apparently rarely produced): female, egg-bearing megaspores;
  and male, sperm-bearing microspores; the 2 types of spores deveoping in separate
  capsules (sporocarps).

NOTE Agriculturally, *Azolla* is important for its symbiosis
with the nitrogen-fixing *Anabaena azollae* Strasburger, a
cyanobacterium (blue-green alga). Because the plants fix
nitrogen, they are often used as a green fertilizer in Asia
and mixed with livestock feed as a nutritional supplement.
Azolla pinnata has been cultivated for many centuries in
rice paddies of Vietnam and China, where it acts as a fertilizer after it decomposes. This species is now reported in
the USA from Florida, Louisiana, and North Carolina.

NAME First described as a genus by Lamarck in 1783; family
name in honor of Italian Antonio Maria Salvini (1633–
1721). *Azolla* from Greek, *azo,* to dry, and *ollupi,* to kill, referring to rapid death of plants when removed from water.

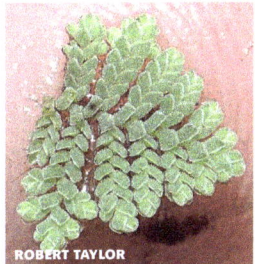

Azolla pinnata

Azolla cristata Kaulfuss
CRESTED MOSQUITO FERN

DISTRIBUTION CT | MA | ME | NH | NJ | NY | PA
FIELD TIPS
- small, mosslike, free-floating, aquatic fern
- may form matlike colonies
- reddish to bright green

HABITAT
Floating on quiet water of river backwaters and ponds; sometimes exposed on wet muddy banks; plants easily broken to form new plants and also dispersed by waterfowl to new habitats.

DESCRIPTION

RHIZOMES Small (less than 3 cm long), pale brown, branching, with tiny roots.

PINNAE Small, less than 1 cm wide; pinnules 2-lobed, the upper smaller lobe floating and lower larger lobe submersed; red when in full sun or under stress, to green when in shade.

SPOROCARPS (rarely seen) Of two kinds, in the leaf axils; the smaller sporocarps contain a single large female megaspore, the larger sporocarps contain many tiny male microspores.

SYNONYMS *Azolla caroliniana* Willd., *Azolla mexicana* Schltdl. & Cham. ex C. Presl, *Azolla microphylla* auct. non Kaulfuss, Evrard and Hove (2004) determined that these former species are best included within *Azolla cristata*.

SIMILAR SPECIES The region's only other true aquatic fern is **water-clover** (*Marsilea*, page 157); however, members of that genus are rooted rather than floating, and the leaf blade is larger and 4-parted.

NOTE The pale bluish green color of the upper lobes is due, in part, to the presence of a symbiotic blue-green bacterium, *Anabaena azollae*.

NAME First described by Kaulfuss in 1824. Common name from supposed ability of dense mats of this plant to smother mosquito larvae.

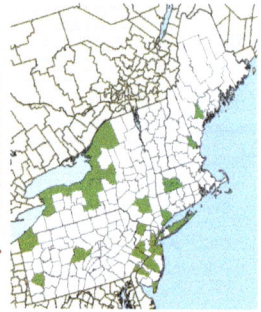

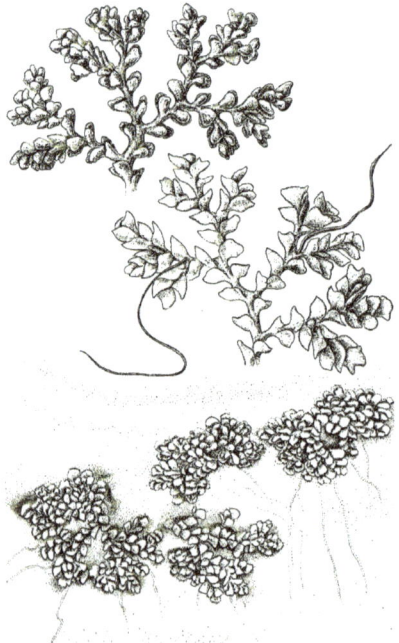

ZIHAO WANG

Azolla cristata CRESTED MOSQUITO FERN

SCHIZAEACEAE *Curly-grass fern family*

TINY FERN; STRONGLY DIMORPHIC, the sterile fronds grass-like and curly; fertile fronds composed of a wiry stipe tipped by a small blade divided into narrow segments. Plants are small and grass-like, and unlikly to be confused with any other of the region's true ferns. Worldwide, two genera and an estimated 35 species; one genus and species in our flora, found along the mid-Atlantic coast (Delaware, New Jersey, southern New York), and in eastern Canada.

Schizaea pusilla Pursh
LITTLE CURLY-GRASS FERN

DISTRIBUTION NJ | NY

FIELD TIPS
• tiny, grass-like plants
• may form matlike colonies
• reddish to bright green

HABITAT

Bogs, wet grassy depressions,sphagnum hummocks overlying wet sand; soils moist and very acidic, often sterile; sometimes under trees of Atlantic white cedar (*Chamaecyparis thyoides*).

DESCRIPTION

RHIZOMES Short, slender, erect, hairy, sending up a clump of tiny fronds to 12 cm tall.

FRONDS Clustered, not fern-like; of 2 forms (fertile and sterile fronds unalike). Sterile fronds linear, 1–6 cm long and to only 1 mm wide, not differentiated into stipe and blade, upright, curled, twisted, wiry; glabrous. Fertile fronds appearing in late summer and fall, 2–12 cm long, stipe much longer than the blade.

FERTILE BLADES 2–4 mm long, pinnately divided into 4–7 pairs of folded, comb-like pinnae.

PINNAE Small, 1.5–4 mm long.

SPORANGIA In 2 rows, with 8–14 per pinna.

CONSERVATION STATUS *endangered* New York.

NOTE Sometimes locally common but difficult to see due to its small size.

NAME First discovered in Burlington County, New Jersey, and named by Pursh in 1814. The genus name is derived from the Greek for *split,* and was proposed by J. E. Smith in 1793. The common name refers to the curled, grass-like, sterile fronds.

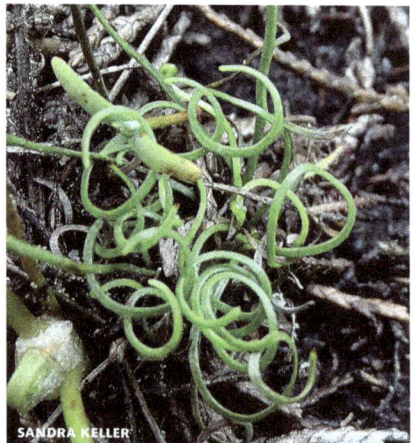

SANDRA KELLER

THELYPTERIDACEAE *Maiden fern family*

MEDIUM-SIZED DECIDUOUS FERNS, spreading by rhizomes to form colonies; sterile and fertile fronds usually alike, 1-pinnate to pinnate-pinnatifid, with transparent needlelike hairs; sori usually on veins (but not marginal) on pinna underside; indusia present and often soon withering, or absent (*Phegopteris*). Worldwide, an estimated 1034 species in 30, mostly tropical genera; 3 genera in our flora.

KEY CHARACTERS
- Deciduous, colony-forming ferns, of forests and sometimes in wetlands.
- Sterile and fertile fronds more or less alike, blades mostly 1–2 pinnate.
- Fronds with needlelike, transparent, usually 1–celled hairs; these also on rhizomes.
- Sori round, indusia present (often soon shed), kidney-shaped; or absent.
- Vascular bundles in stipe 2, these joined upwards.

NOTE Taxonomic questions remain for this family. Our species previously included in genus *Thelypteris,* but are here separated into 3 genera; check synonyms listed for each species to see alternate scientific names.

NAME Family name from *Thelypteris,* Greek *thelys,* female, and *pteris,* fern.

KEY TO THELYPTERIDACEAE (ALL GENERA AND SPECIES)

1 Leaf blades 7–25 (–30) cm long, triangular, not more than 2x as long as wide; rachis with wings between the pinnae; sori without indusia; midribs of pinnae lacking a groove on upper surface **2**

1 Leaf blades (15-) 20–100 cm long, lance-shaped, oblong-lance-shaped, or triangular, more than 2x longer than wide; rachis without wings between the pinnae; sori with kidney-shaped indusia; midribs of pinnae with a groove on upper surface **4**

2 Rachis winged between all pinnae pairs; stipes becoming smooth just below base of blade (within 1 cm), apart from a few scales; blades usually wider than long; regionwide . . . . . .
. . . . *Phegopteris hexagonoptera*, BROAD BEECH FERN, page 232

2 Rachis not winged between lowest two pinnae pairs; stipes finely hairy and usually also scaly in lower portion; blades usually longer than wide; regionwide. **3**

3 Colonies fairly compact, often dense with overlapping fronds; fronds shorter; lowermost pinnae relatively wide (ca. 4× longer than wide) but never as wide as in *Phegopteris hexagonoptera;* often strongly deflexed and projecting forward from the plane of the upper frond; the overall shape ovate with a long-acuminate tip; regionwide . . . .
. . . . *Phegopteris connectilis*, NARROW BEECH FERN, page 230

3 Colonies often loose, fronds touching but not overlapping; fronds taller; lowermost pinnae relatively narrow (4.5× or 5× longer than wide); somewhat but not strongly deflexed and not projecting as far forward; overall blade shape deltate (triangular, but with a long tip) and less rounded at bottom; CT, MA, ME, NH, NY, VT . . . .
. . . . *Phegopteris excelsior*, TALL BEECH FERN, page 230

4 Leaf blade broadest near middle, gradually reduced to base; stipe less than 1/3 length of blade; plants of upland and wetland habitats; common, regionwide . . . .
. . . . *Amauropelta noveboracensis*, NEW YORK FERN, page 226

4 Leaf blade broadest near base, the pinnae stopping abruptly; stipe 2/3 to fully as long as blade; plants of wetlands **5**

5   Underside of blade without glands; lateral veins of sterile lobes once-forked between pinnule midvein and margin; lower surface of pinna midrib (costa) with tan, ovate scales; lobes of fertile leaves inrolled; indusia divided into hairlike segments (rarely smooth); common, regionwide .... *Thelypteris palustris*, EASTERN MARSH FERN, page 234

5   Underside of blade with small, stalkless, globular, golden to reddish glands; lateral veins of sterile lobes simple, not forked between the pinnule midvein and the margin; lower surface of pinna midrib lacking scales; lobes of fertile leaves flat to slightly revolute; indusia with minute glands along the margins; regionwide............. *Coryphopteris simulata*
.... MASSACHUSETTS FERN, page 228

Amauropelta Kunze NEW YORK FERN

WORLDWIDE, about 215 species in the genus; one in our region.

Amauropelta noveboracensis (L.) S.E. Fawc. & A.R. Sm.
NEW YORK FERN

DISTRIBUTION CT | MA | ME | NH | NJ | NY | PA | RI | VT
FIELD TIPS
• medium, deciduous, colony-forming fern
• fronds yellow-green, in clumps of several from rhizome
• blade tapering to both ends, lowest pinnae very small, near base of frond

HABITAT
Moist shady woods, especially near swamps, streams, and seeps in ravines, often in slightly disturbed secondary forests, may form dense colonies, especially in openings; soils moist, moderately acidic.

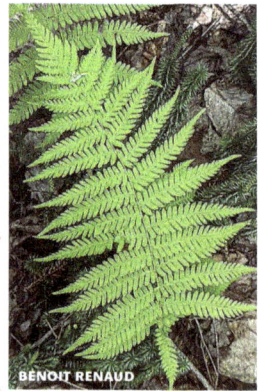
BENOIT RENAUD

DESCRIPTION

ROOTSTOCK Long-creeping, cord-like, branching; scales few or absent.

FROND Deciduous, in tufts of 3 or more fronds along the rhizome; sterile and fertile fronds somewhat different; the fertile fronds larger, more erect and narrower than sterile fronds; to about 70 cm long by 15 cm wide; becoming whitish when dying.

STIPE Straw-colored, shading to green; scales tan to reddish brown; vascular bundles 2 at stipe base, merging above into a U-shape.

RACHIS Pale green to straw-colored, covered with whitish hair.

BLADE Elliptic, tapering to both ends, lowest pinnae very small (less than 1 cm long) and near base of frond; 1-pinnate-pinnatifid, the pinnule divisions falling short of

meeting the costa by 1 mm; yellow-green; underside with transparent needlelike hairs.

PINNAE 20 to 24 pairs, lance-shaped, long tapering, less cut near the ends; costae grooved above, discontinuous with the rachis; margins entire or crenate, hairy; veins free, forked or not.

SORI Round, small, nearer the margin than the costule, absent from ends of the pinnules (or pinnae); indusium kidney-shaped, tan, withering.

SYNONYMS *Dryopteris noveboracensis* (L.) A. Gray, *Parathelypteris noveboracensis* (L.) Ching, *Polypodium noveboracense* L., *Thelypteris noveboracensis* (L.) Nieuwl., *Thelypteris thelypterioides* (Michx.) Holub

SIMILAR SPECIES **Hay-scented fern** (*Dennstaedtia punctilobula*) is larger, hairy, and with fronds mostly single in lines along rhizome; lowest pinnae larger and higher on frond; **eastern marsh fern** (*Thelypteris palustris*) usually found in wetter habitats.

NOTE Sent to Linnaeus from "Canada" (probably ne USA) and named by him *Polypodium noveboracense* in 1753. Transferred to *Thelypteris* by Nieuwland in 1910.

Amauropelta noveboracensis **NEW YORK FERN**

ELIAS

Coryphopteris Holttum BOG FERN

A GENUS OF 47 SPECIES WORLDWIDE, one in our flora.

Coryphopteris simulata (Davenp.) S.E. Fawc.
MASSACHUSETTS FERN

DISTRIBUTION CT | MA | ME | NH | NJ | NY | PA | RI | VT

FIELD TIPS
• medium-size deciduous fern of wet, acidic habitats
• pinnae narrow, separated along rachis

HABITAT
Acidic swamps and bogs, often growing with sphagnum mosses; in moist to wet shaded woodlands; peaty meadows.

DESCRIPTION

ROOTSTOCK Long-creeping, branching, some stipes distantly spaced, some closely spaced; scales sparse, pale brown.

FROND Deciduous, to 80 cm long by 15 cm wide; sterile and fertile fronds nearly alike, though sterile fronds arching, and fertile fronds taller, more erect, with longer stipes.

STIPE Brown at base, shading to yellow-green, with a few tan scales near base; vascular bundles 2, merging above.

RACHIS Green; slightly hairy on upperside, smooth below.

BLADE Lance-shaped, 1-pinnate-pinnatifid; pinnae widely separated, tapering to the tip, widest at or just above lowest pinna, lowest pair of pinnae bending down and forward; pale green; with short, whitish hairs, these soon falling, and with reddish to orangish, resinous, shiny, round glands.

PINNAE 16 to 18 pairs, lance-shaped, long tapering, less incised near ends, sessile; lower pinnae narrowed near rachis; fertile pinnules slightly inrolled; costae grooved above, discontinuous with the rachis; margins entire; veins free, simple.

SORI Round, midway between the margin than the costule, merging into each other; indusium narrowly kidney-shaped, whitish, maturing to tan, with tiny orange glands (use hand lens to see); sporangia brown.

SIMILAR SPECIES New York fern (*Amauropelta noveboracensis*), but lowest pinnae in that species are very small; separated from **eastern marsh fern** (*Thelypteris palustris*) by the veins never forking, the presence of small glands on the pinna underside, and the bending forward and down of the low-

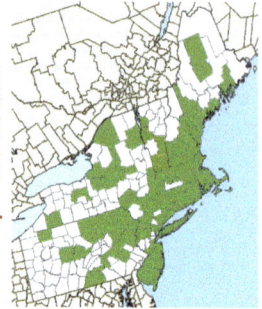

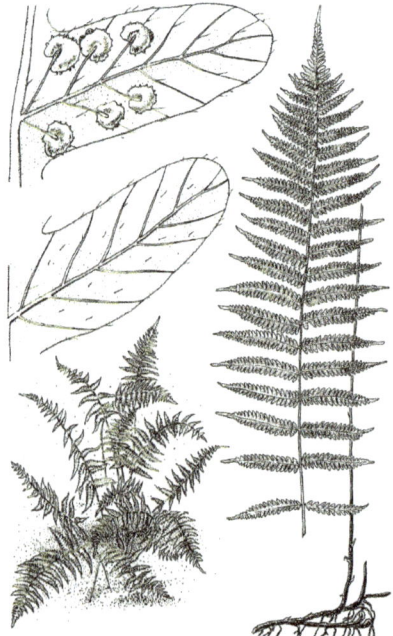

est pinna pair; in eastern marsh fern, the veins often forked and the glands absent.

SYNONYMS *Dryopteris simulata* Davenport, *Parathelypteris simulata* (Davenport) Holttum, *Thelypteris simulata* (Davenport) Nieuwl.

NAME First collected by Dodge in New Hampshire, in 1880, and named by Davenport *Aspidium simulatum* in 1894. Transferred to *Thelypteris* by Nieuwland in 1910. 'Simulata' means resembling, in this case, perhaps to forms of lady fern (*Athyrium* spp.), *Amauropelta noveboracensis*, and *Thelypteris palustris*.

JOHN BAUR

MICHAEL ELLIS

Coryphoteris simulata MASSACHUSETTS FERN

Phegopteris Fée **BEECH FERN**

WORLDWIDE, 4 SPECIES REPORTED FOR THE GENUS; 3 species in our region.

ADDITIONAL SPECIES **Tall beech fern** (*Phegopteris excelsior* N.R. Patel & A.V. Gilman) a newly defined species, is reported from scattered locations in the region (except absent from New Jersey, Pennsylvania, and Rhode Island (Patel et al., 2019); see family key, page 225.

Phegopteris connectilis (Michx.) Watt
NARROW BEECH FERN

DISTRIBUTION CT | MA | ME | NH | NJ | NY | VT

FIELD TIPS
- deciduous creeping fern of moist woods and rocky places, may form colonies
- blade triangular, longer than wide
- lowest pinnae drooping down, not connected to other pinnae by wing along rachis
- indusia absent

HABITAT

Shaded moist to wet forests, streambanks, rocky ravines; sometimes on cliffs and near waterfalls.

DESCRIPTION

ROOTSTOCK Long-creeping, branching, cord-like, densely hairy and scaly when young, becoming smoother with age.

FROND Deciduous, new fronds all summer, sterile and fertile fronds alike, arching, to 50 cm long by 15 cm wide.

STIPE Straw-colored at base, with tan, lance-shaped scales and pointed hairs; vascular bundles 2, elongate, sometimes joined at top of stipe.

RACHIS Green, with transparent needlelike hairs and spreading, lance-shaped scales, not winged between lowest pinnae.

BLADE Triangular, widest at base, usually somewhat longer than wide, 1-pinnate-pinnatifid; light green, often finely hairy on both sides.

PINNAE 12 to 15 pairs, ovate-lance-shaped, base fused to rachis, except lowest pair unconnected by wings to the upper ones, and also typically drooping down and out; costae not grooved; veins free, simple or forked.

SORI Round, near margin; indusium absent; sporangia tan.

SYNONYMS *Dryopteris phegopteris* (L.) C. Chr., *Lastrea phegopteris* (L.) Bory, *Phe-*

gopteris polypodioides Fée, *Thelypteris phegopteris* (L.) Slosson

CONSERVATION STATUS *threatened* Rhode Island.

SIMILAR SPECIES **Broad beech fern** (*Phegopteris hexagonoptera*) has a winged rachis between the 2 lowermost pinnae pairs.

NAME Known to Linnaeus in Europe, and named by him *Polypodium dryopteris* in 1753; included in *Phegopteris* by Fée nearly 100 years later. Genus name from *phegos*, beech, and *pteris*, fern, in reference to its growing under beech trees. Epithet from Latin, *conecto*, to fasten together, referring to the upper joined pinnae.

Phegopteris connectilis NARROW BEECH FERN

Phegopteris hexagonoptera (Michx.) Fée
BROAD BEECH FERN

DISTRIBUTION CT | MA | ME | NH | NJ | NY | PA | RI | VT

FIELD TIPS
- deciduous fern of moist woods, forming small colonies
- blade about as wide as long
- all pinnae pairs connected by wing along rachis, lowest pair largest, spreading
- indusia absent

HABITAT
Rich, moist deciduous forests, often at bases of slopes, edges of seeps, and along streams; soils moderately acidic to alkaline.

DESCRIPTION

ROOTSTOCK Long-creeping, branching, cord-like, densely scaly and with a few hairs.

FROND Deciduous, arching, sterile and fertile fronds alike, to about 60 cm long by 30 cm wide.

STIPE Reddish-brown at base, straw-colored above; scales tan, lance- shaped, and sometimes with hairs, vascular bundles 2, elongate, at 90°, merging or not by the top of the stipe to an open U-shape.

RACHIS Green to straw-colored, with whitish scales; winged throughout.

BLADE Broadly triangular, usually as wide as long, 1-pinnate-pinnatifid above to 2-pinnate-pinnatifid below; light green; underside with hairs along costae and veins, also with yellowish stalked glands.

PINNAE 12 to 15 pairs, ovate-lance-shaped, all connected by wing along rachis (including lowermost pinnae pair); basal pinnae largest, more spreading and more dissected than upper pinnae; costae not grooved; margins lobed, hairy; veins free, simple or forked.

SORI Round, sparse, near end of a vein; indusium absent; sporangia tan.

SYNONYMS *Dryopteris hexagonoptera* (Michx.) C. Chr., *Thelypteris hexagonoptera* (Michx.) Weath.

SIMILAR SPECIES **Narrow beech fern** (*Phegopteris connectilis*) lacks the winged rachis between the 2 lowermost pinnae pairs.

NAME Though early collected in Virginia and figured by Plukenet in 1691, this was confused with *P. connectilis* by Linnaeus.

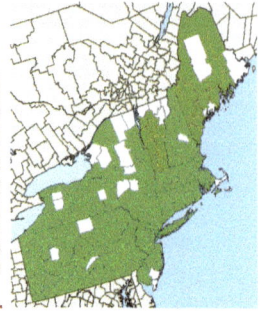

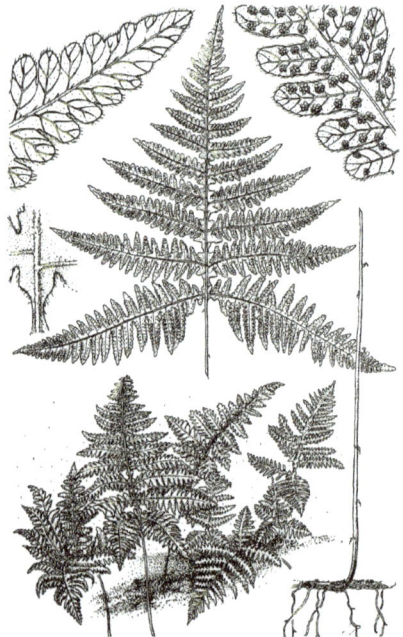

Michaux named it *Polypodium hexa-gonopterum* in 1803, and Fée placed it in his genus *Phegopteris* in about 1850. 'Hex' means six; *gona*, joint; *pterus* from *pteron,* Greek for wing, thus a six-jointed wing, referring to the angular appearance of the winged rachis.

RYAN SORRELLS

MATT BERGER

RYAN SORRELLS

Phegopteris hexagonoptera BROAD BEECH FERN

Thelypteris Schmidel **MARSH FERN**

WORLDWIDE, TWO SPECIES REPORTED, one species in our flora.

Thelypteris palustris Schott
EASTERN MARSH FERN

DISTRIBUTION CT | MA | ME | NH | NJ | NY | PA | RI | VT
FIELD TIPS
• common deciduous fern of wet places
• blade widest below middle
• stipe long, brown
• veins once-forked

HABITAT
Wet meadows, swamps, wet thickets, ditches, fens, bogs, wet
shores, seeps, etc., in sun or partial shade, often on raised
hummocks or sometimes in standing water.

DESCRIPTION
ROOTSTOCK Long-creeping, branching, black behind the
white tip; scales sparse.

FROND Deciduous, often twisted; sterile and fertile fronds
somewhat different; sterile fronds earlier in season, fertile
fronds narrower, taller, developing about 6 weeks later, to
75 cm long by 20 cm wide.

STIPE Usually longer than blade, straw-colored or darker
near base, shading upward to green, with a few scales near
base, vascular bundles 2, oblong at stipe base, merging
above to an open u-shape.

RACHIS Pale green to purplish green;
smooth and slender; grooved.

BLADE Lance-shaped, 1-pinnate-pinnatifid,
pinnae widely separated, tapering to tip,
widest below the middle, narrowing
somewhat to the lowest pinna, thin and
delicate, pale green; with short, whitish
hairs, these falling soon.

PINNAE 14 to 20 pairs, lance-shaped, long
tapering, less dissected near the ends,
sometimes twisted toward horizontal, sessile; costae grooved above, discontinuous
with rachis; margins of sterile pinnae entire to slightly toothed; fertile pinnae triangular, margins folded downward to
cover some of the sori; veins free, once-
forked.

SORI Round, midway between the margin
than the costule, merging into each other;
indusium small, irregular or kidney-
shaped, often hairy but without glands,
soon shed.

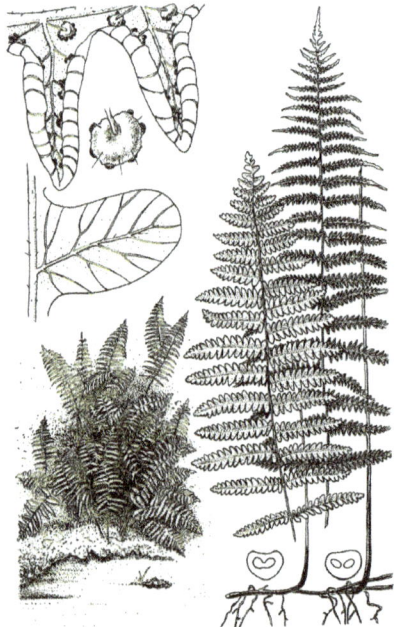

SIMILAR SPECIES
- **Massachusetts fern** (*Coryphopteris simulata*) similar but all veins unforked vs. 1–forked in *Thelypteris palustris*.
- **Hay-scented fern** (*Dennstaedtia punctilobula*) has hairy stipes shorter than the blade.
- **Lady fern** (*Athyrium* spp.) is clumped, with true pinnules and toothed margins.

NAME Linnaeus in 1753 gave the name *Acrostichum thelypteris* to a European fern, and 9 years later Schmidel made *Thelypteris* a genus. Schott proposed *T. palustris* in 1834. In 1848 Gray made the combination *Dryopteris thelypteris*. Greek *thelys*, female, and *pteris*, fern; 'palustris' means marshy or swampy.

MARINA SADYKOVA   fertile frond

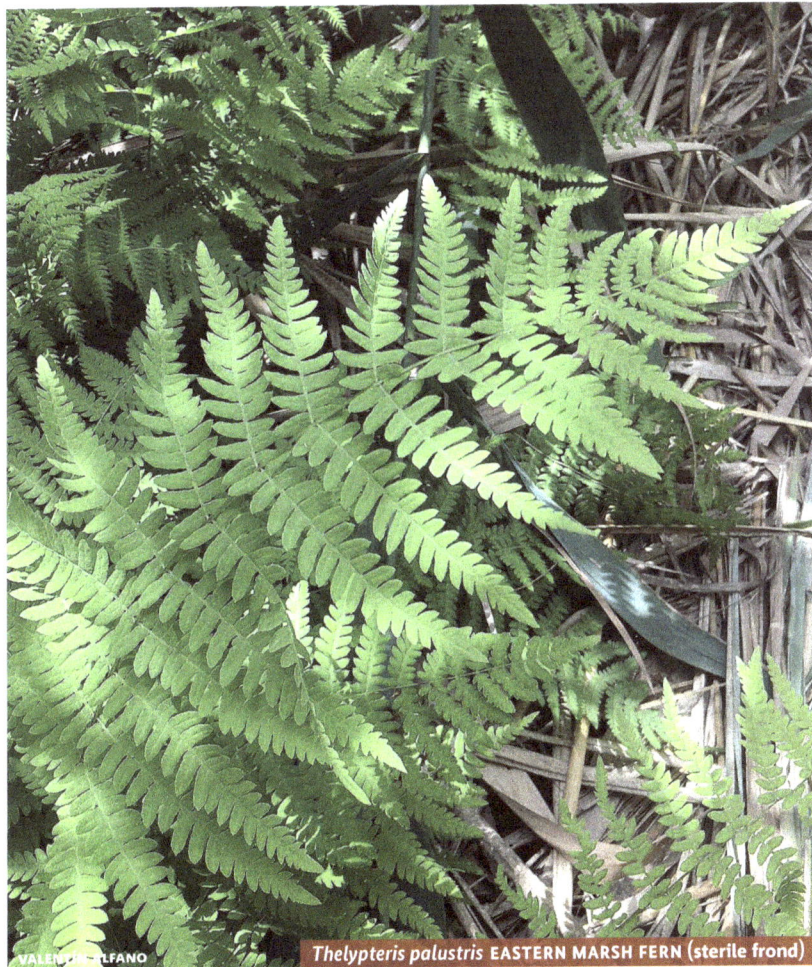

VALENCIA ALFANO

Thelypteris palustris EASTERN MARSH FERN (sterile frond)

WOODSIACEAE   *Cliff fern family*

Woodsia R. Br.   CLIFF FERN

SMALL DECIDUOUS FERNS, mostly growing on rock cliffs and ledges; sterile and fertile fronds alike. Plants generally covered with hairs, scales and glands; also, in some species, the stipes are jointed, the frond eventually breaking at this point, and leaving a more or less even stubble. Worldwide, the family contains the single genus *Woodsia*, and an estimated 39 species found in northern temperate regions and at high-elevations throughout the tropics; 4 species in our flora.

KEY CHARACTERS
- Fronds covered with hairs, scales or glands
- Stipes jointed in some species, the frond breaking at the joint to leave a more or less even stubble
- Veins ending before margin, sometimes the vein tip swollen.
- Indusium cup-like or dissected into several straplike or scalelike segments which encircle the round sorus.

NAME Named for Joseph Woods, 1776–1864, an English botanical author.

SIMILAR SPECIES *Woodsia* and *Cystopteris* are small ferns with thin-textured leaves, often growing together on or near rock outcrops, and often confused. *Woodsia* has the indusium divided into a series of scalelike or hairlike structures, attached below the sorus; *Cystopteris* has an undivided indusium, pocket-like or hood-like, attached around one side of the sorus. *Woodsia* has persistent dark stipe bases; in *Cystopteris* the stipe bases are deciduous; *Woodsia* stipes are scaly most of the way to the blade; *Cystopteris* stipes are without scales except near the stipe-base. In *Woodsia*, veins do not reach the margin and tip of vein often enlarged; *Cystopteris* veins extend to margin.

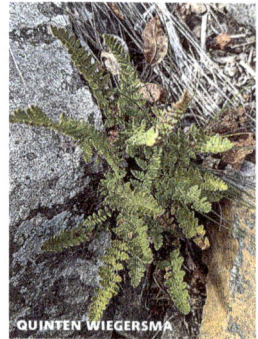

QUINTEN WIEGERSMA

Woodsia ilvensis
RUSTY CLIFF FERN

HYBRIDS *Woodsia × gracilis* (Lawson) Butters, hybrid between *W. alpina* and *W. ilvensis;* known from Vermont.

KEY TO *WOODSIA* | CLIFF FERN

1 Stipe jointed above base, the old stubble more or less uniform height 2
1 Stipe not jointed, the stubble irregular; regionwide. . . . . *Woodsia obtusa*
. . . . BLUNT-LOBE CLIFF FERN, page 244
2 Fronds delicate, smooth; stipe green; blade to only 1.5 cm wide, lower pinnae reduced to wings; indusium splitting into short hairs; rare in MA, ME, NH, NY, VT . *Woodsia glabella*
. . . . SMOOTH CLIFF FERN, page 240
2 Fronds firm; stipe and lower rachis brown; indusium splitting into many long hairs. 3
3 Blade narrow, nearly smooth; rare in ME, NY, VT . . . . *Woodsia alpina*
. . . . NORTHERN CLIFF FERN, page 238
3 Blade wider, densely hairy; regionwide but absent from RI . . . . *Woodsia ilvensis*
. . . . RUSTY CLIFF FERN, page 242

BONNIE SEMMLING

Woodsia obtusa BLUNT-LOBE CLIFF FERN

Woodsia alpina (Bolton) S.F. Gray
NORTHERN CLIFF FERN

DISTRIBUTION ME | NY | VT

FIELD TIPS
- very small, erect, clumped fern
- stipes dark brown or purple when mature, jointed, breaking to leave uniform stubble
- sori near margins, indusia of hairlike segments
- acidic rock habitats

HABITAT

Crevices and ledges of moist, partially shaded cliffs of acidic rock.

DESCRIPTION

ROOTSTOCK Erect, compact, with numerous persistent stipe bases of nearly equal length; scales uniformly brown.

FROND Deciduous, in dense clusters; sterile and fertile fronds alike; 10–20 cm long.

STIPE Persistent base, red-brown or dark purple when mature, jointed above base at swollen node halfway up stipe, red-brown lance-shaped scales at base, fewer upwards; vascular bundles 2.

RACHIS With scattered scales and red-brown hairs.

BLADE Lance-shaped, broadest below middle; 1-pinnate-pinnatifid (barely so), bright green.

PINNAE 8 to 15 pairs, largest pinnae with 1–3 pairs of pinnules, the shorter ones merely fan-shaped; costae grooved above, grooves continuous from rachis to costae; margins nearly entire, merely lobed; veins free, simple or forked.

SORI Round, near the margin; indusium dissected into hairlike segments enveloping sorus, unravelling with maturity; sporangia brownish.

SYNONYM *Woodsia glabella* var. *bellii* (G. Lawson) G. Lawson

CONSERVATION STATUS *endangered* New York, Vermont; *threatened* Maine.

SIMILAR SPECIES Distinguished from **rusty cliff fern** (*Woodsia ilvensis*) by having scattered hairs and scales, and the largest pinnae with only 1–3 pairs of pinnules; W. ilvensis has abundant hairs and scales; its largest pinnae have 4–9 pairs of pinnules; also a much more common species in our region.
- **Smooth cliff fern** (*Woodsia glabella*) by having a dark, hairy stipe.

NOTE Originated as a hybrid between *Woodsia ilvensis* and *Woodsia glabella*.

NAME Fiirst reference to this species was in John Ray's 1690 *Synopsis,* which recorded the discovery of a rare fern near the summit of Snowdon in Wales; however, the genus *Woodsia* was not established until 1810 by Robert Brown, who named it after Joseph Woods, an English botanist.

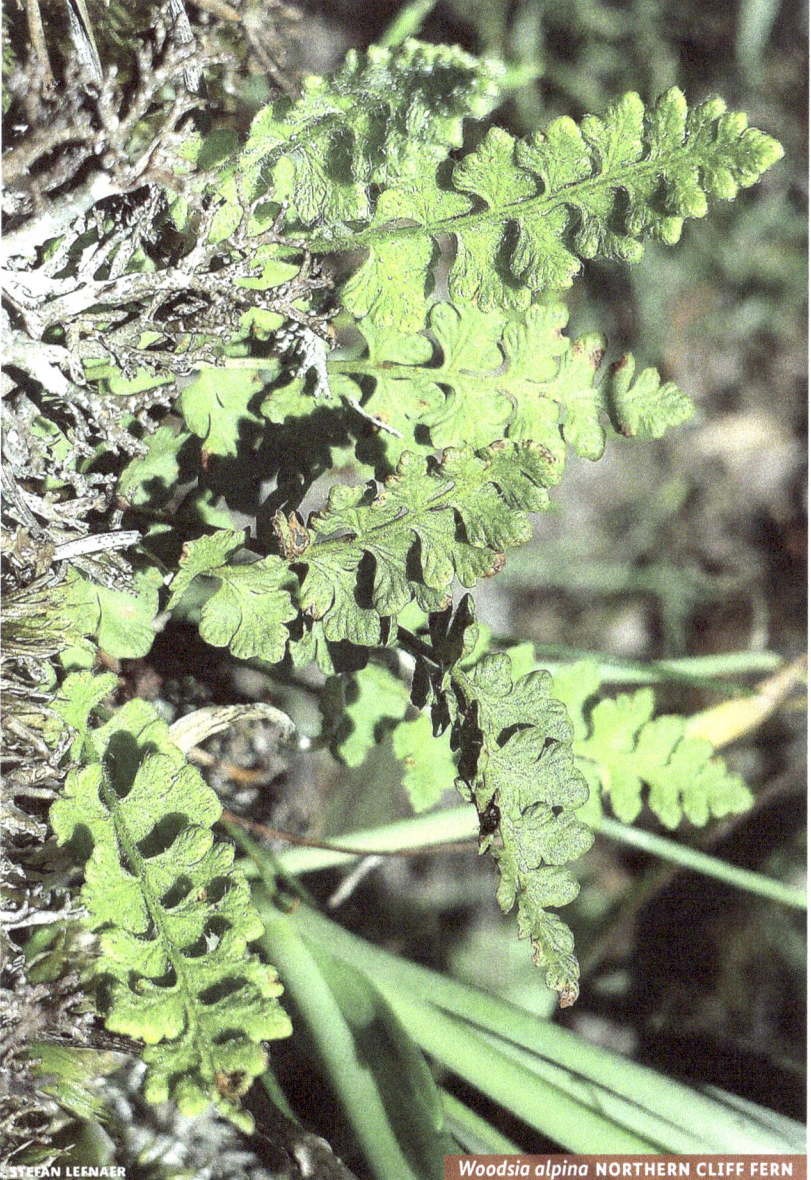

Woodsia alpina NORTHERN CLIFF FERN

Woodsia glabella R. Br. ex Richards.
SMOOTH CLIFF FERN

DISTRIBUTION MA | ME | NH | NY | VT

FIELD TIPS
- very small, deciduous, clumped fern
- blades pale green, without hairs or scales
- stipes green to straw-colored, jointed
- fern of rock crevices

HABITAT

Crevices in moist, north-facing cliffs, the rock basic to acidic.

DESCRIPTION

ROOTSTOCK Erect, compact, with abundant persistent petiole bases of nearly equal length; scales uniformly brown. Frond Deciduous, in small clusters; sterile and fertile fronds alike; 5–15 cm long.

STIPE Green or straw-colored throughout, jointed above base at swollen node, to only ca. 3 cm long; scales broadly lance-shaped, to 4 mm, yellow-brown at base, smooth above the joint; vascular bundles 2, round or oblong.

RACHIS Green, smooth.

BLADE Linear to linear-lance-shaped, 1-pinnate-pinnatifid; pale green, without hairs.

PINNAE 7 to 9 pairs, sessile, largest pinnae with 1–3 pairs of pinnules, lower pinnae sometimes fan-shaped; costae grooved above, grooves continuous from rachis to costae; margins entire or crenate, often folding downwards; veins free, simple or forked.

SORI Small, near margin; indusium a saucer-shaped disk underneath sori, dissected into hairlike segments, these visible only while sori young; sporangia brownish.

CONSERVATION STATUS *endangered* New Hampshire Massachusetts, New York; *threatened* Maine, Vermont.

SIMILAR SPECIES Distinguished from region's other *Woodsia* by its yellow-green stipe and lack of hairs. **Bright-green spleenwort** (*Asplenium viride*) has elongate sori, the indusium attached to side of sorus and not below it.

NAME Latin, diminutive form of *glaber,* smooth or without hair.

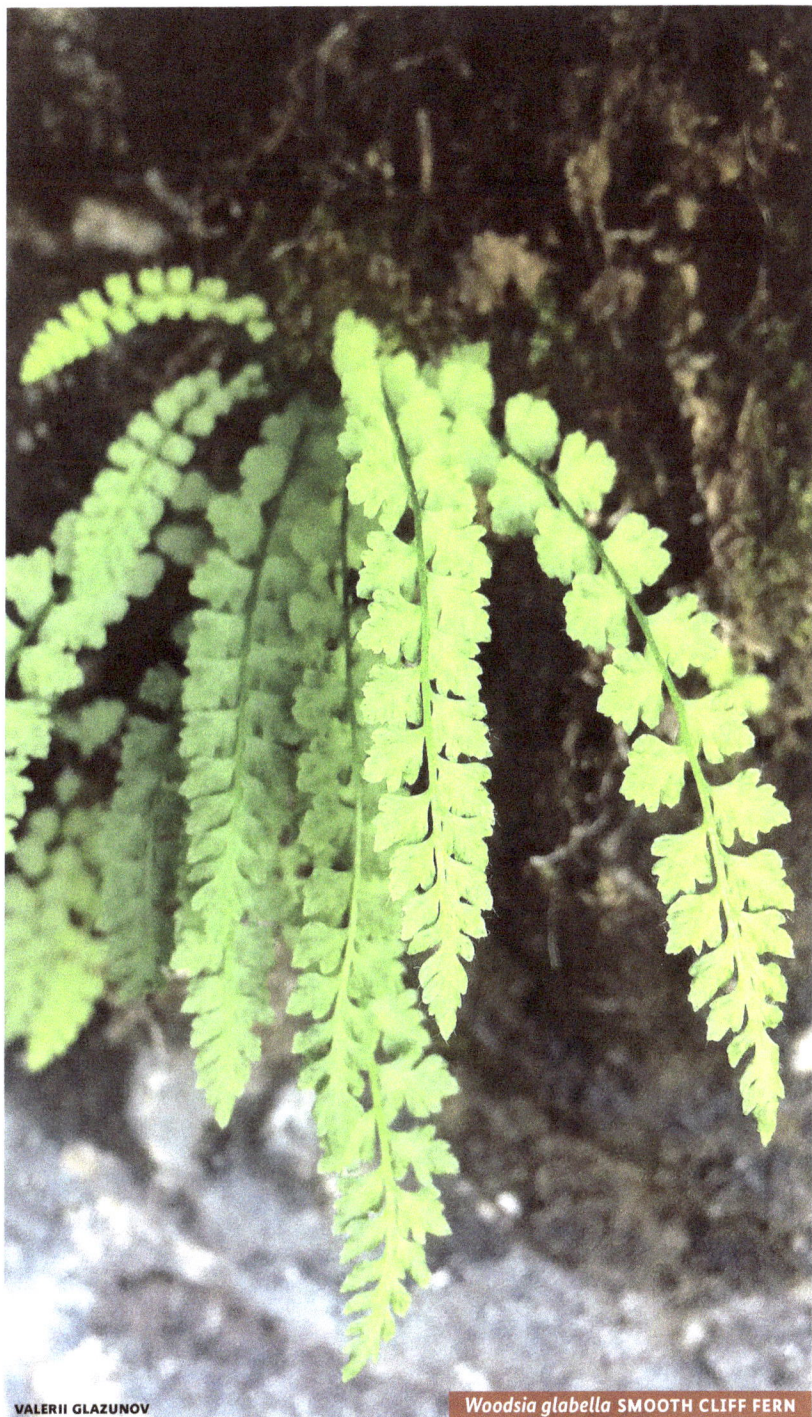

VALERII GLAZUNOV

Woodsia glabella SMOOTH CLIFF FERN

Woodsia ilvensis (L.) R. Br.
RUSTY CLIFF FERN

DISTRIBUTION CT | MA | ME | NH | NJ | NY | PA | RI *historical* | VT

FIELD TIPS
- small, deciduous, clumped fern of acidic rock
- rachis and blade underside densely hairy, the hairs white or rusty brown
- mature stipe dark brown or purple, jointed above base

HABITAT

Dry, exposed or partly shaded cliffs and ledges of non-calcareous rock

DESCRIPTION

ROOTSTOCK Short, often branched, dark brown to black, with abundant persistent petiole bases of nearly equal length; scales many, uniformly brown.

CROZIER Covered with silvery white hairs, may be present all summer.

FROND Deciduous, in dense clusters; sterile and fertile fronds alike; 5–25 cm long.

STIPE Persistent base, brown or dark purple when mature, jointed above base at swollen node halfway up stipe; red-brown lance-shaped scales near base, a mixture of scales and hairs above and into the rachis; vascular bundles 2, oblong at stipe base, merging above to an open U-shape.

RACHIS Green, densely hairy and scaly.

BLADE Lance-shaped, broadest below middle, 2-pinnate at base, less divided above; underside with dense silvery-gray hairs when young, turning rusty-brown later, with narrow scales below, and multicellular hairs along costa above.

PINNAE 10 to 20 pairs, lowest somewhat smaller, sessile or nearly so; with 4–9 pairs of rounded, variable lobes; costae grooved above, grooves continuous from rachis to costae; margins hairy, entire or crenate, often folding downwards; veins free, simple or forked.

SORI Small, near margin, often hidden by dense hairs and scales; indusium fringed with hairlike segments enveloping sorus, persistent but often obscure; sporangia brownish.

SIMILAR SPECIES Distinguished from **Northern cliff fern** (*Woodsia alpina*) by having abundant hairs and scales, and the greater number of pinnules (4–9 pairs on

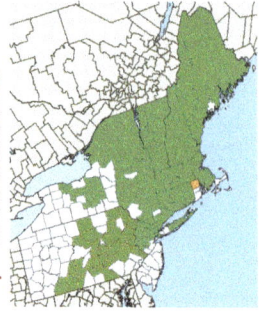

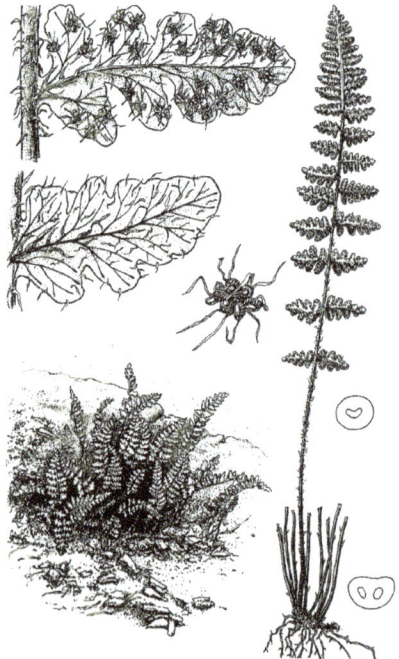

each pinna); **smooth cliff fern** (*Woodsia glabella*) by having hairs and scales.

NOTE Plants may dry and turn brown during drought, with
green fronds formed following rain.

NAME In 1753, Linnaeus named *Acrostichum ilvense,* a fern
found in parts of Europe (the species name referring to the
island of Elba). In 1810, Robert Brown placed these ferns in
a new genus he named *Woodia,* correcting it to *Woodsia*
five years later. Plants from America proved so similar to
the European that the same names were applied.

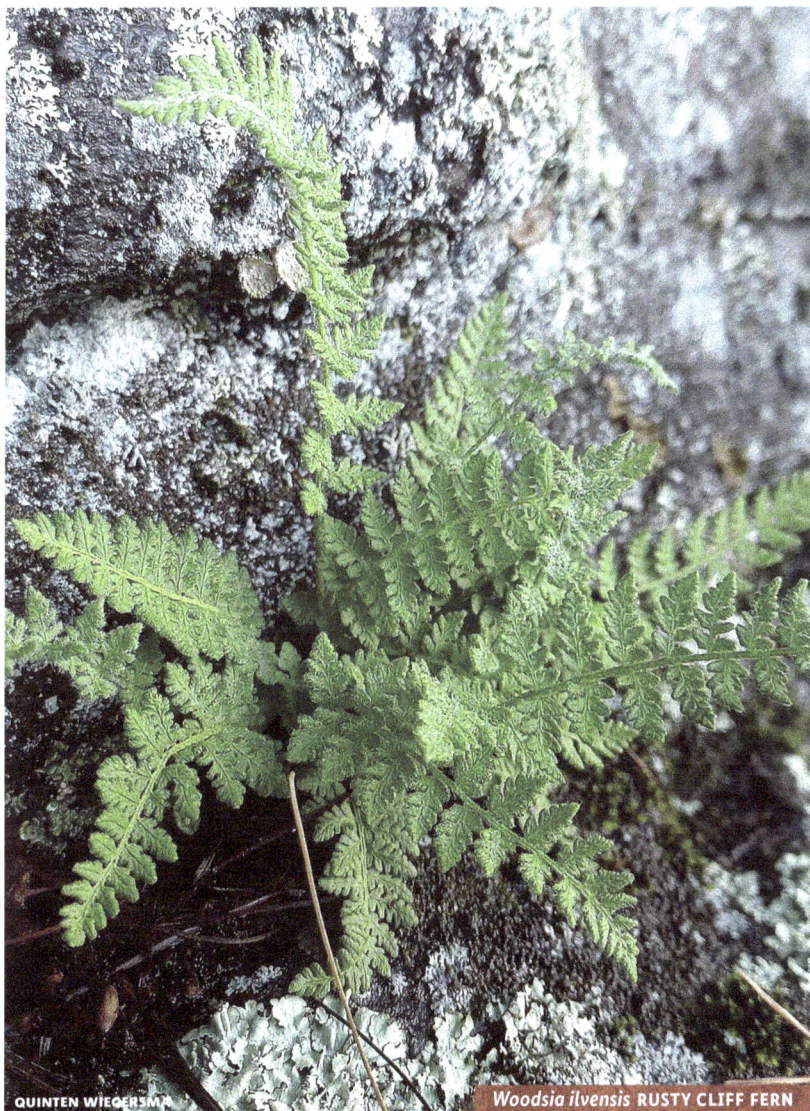

QUINTEN WIEGERSMA   *Woodsia ilvensis* RUSTY CLIFF FERN

Woodsia obtusa (Spreng.) Torr.
BLUNT-LOBE CLIFF FERN

DISTRIBUTION CT | MA | ME | NH | NJ | NY | PA | RI | VT

FIELD TIPS
- nearly evergreen, loosely clumped fern, usually on calcareous rock
- stipe bases persistent, of unequal length, straw-colored to light brown when mature
- blade with glands on both sides
- region's largest cliff fern

HABITAT

Open to partly shaded, outcrops and cliff ledges; the rock often (but not always) calcareous.

DESCRIPTION

ROOTSTOCK Erect to short-creeping, often branched, dark-brown to black, with stipe bases of unequal length and with persistent old fronds attached; scales few.

FROND Nearly evergreen, clustered; old fronds persistent; sterile and fertile fronds nearly alike, but the sterile fronds decline and remain green through winter; 25–40 cm long.

STIPE Not jointed, the persistent bases of unequal lengths, light brown or straw-colored when mature, occasionally darker at base, scales tan some with darker central stripe; vascular bundles 2, oblong at stipe base, merging above to a U-shape.

RACHIS Pale yellow to brown, with glandular hairs and scattered, narrow scales.

BLADE Tapering to both ends, truncated and 2-pinnate-pinnatifid at base; herbaceous, gray-green, glands on both surfaces.

PINNAE 8 to 15 separated pairs, held nearly horizontally, lowest somewhat reduced, sessile or nearly so, pairs further apart the closer to the base; pinnules 3–8 pairs before changing to a pinnatifid tip; glandular hairs on both upper and lower surface and on costae; costae grooved above, grooves continuous from rachis to costae; margins dentate or lobed; veins free, simple or forked, ending before the margin, vein tips usually enlarged.

SORI Round, in one row between midrib and margin; indusium starlike, with 4–6 translucent lobes encircling sorus; sporangia brown, then black.

CONSERVATION STATUS *endangered* New Hampshire, *threatened* Maine.

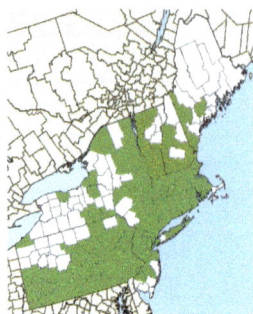

SIMILAR SPECIES Distinguished from **brittle bladder fern**
(*Cystopteris fragilis*) by pinnae with glandular hairs, and
the starlike indusium under sori.

NAME Discovered by Muhlenberg in Pennsylvania, and
named by Sprengel in 1804 *Polypodium obtusum,* which
was changed by Torrey in 1840 to *Woodsia obtusa;* 'obtusa'
refers to the blunt pinnae.

SANDY WOLKENBERG

Woodsia obtusa BLUNT-LOBE CLIFF FERN

ISOETACEAE *Quillwort family*
Isoetes L. QUILLWORT

WORLDWIDE, THE QUILLWORT FAMILY INCLUDES A SINGLE GENUS, *ISOETES,* of about 250 species, with nearly 40 species present in North America, and 9 species in our flora. *Isoetes* are small, perennial, usually aquatic herbs. Plants often totally submerged or sometimes in drier situations along shores. Leaves grasslike, with internal air chambers, broadened and flattened at base, in clusters of few to many from a swollen, cormlike rhizome with numerous long, fleshy roots. Innermost leaves usually sterile; the next outer leaves usually with a small pocket at their expanded base containing the sporangia. The sporangia more or less covered by the thin edges of the velum, and with a small ligule situated above.

As in the spikemosses (*Selaginella*) and water ferns (*Azolla, Marsilea*), spores of *Isoetes* are of two types, numerous, and the surface is variously patterned; relatively large white megaspores (female) are borne in megasporangia (the case containing the spores); smaller microspores (male) are borne in microsporangia.

KEY CHARACTERS
- Small aquatic or wetland plants of lakes and shores.
- Leaves tufted, grass- or chive-like, swollen at base.
- Sporangia in a pocket at base of leaves.

NAME From Greek, *isos,* equal, and *etos,* year, referring to evergreen habit of some species. The long pointed, hollow leaves suggest a quill; 'wort' is an old English word to refer to any herbaceous plant.

NOTE Identification of species is dependent on characters of the mature megaspores; these are about 0.5 mm wide, and a 20× hand lens (or more powerful microscope) may be needed. In the field, surface features are best seen in August or later and when the megaspore is dry. Quillworts can be distinguished from similar aquatic grasslike plants by the swollen leaf base which contains the megaspores.

KEY TO *ISOETES* | QUILLWORT

1 Megaspores conspicuously covered with small spines; regionwide .. *Isoetes echinospora*
.... SPINY-SPORED QUILLWORT, page 248
1 Megaspores not spiny, variously ornamented; leaves persisting more than 1 year....... 2
2 Megaspores averaging less than 0.5 mm wide (ranging from 0.4–0.55 mm), patterned with an unbroken reticulum (net-like covering); regionwide............. *Isoetes engelmannii*
.... ENGELMANN'S QUILLWORT, page 249
2 Megaspores averaging more than 0.5 mm wide (ranging from 0.4–0.75 mm), cristate, rugulate, or reticulate with broken or anastomosing ridges................................ 3
3 Leaves with abundant stomates, bright green; plants aquatic to amphibious, sometimes emergent on shores; girdle of megaspores obscure; sporadic regionwide but absent from ME and NH .... *Isoetes riparia*, SHORE QUILLWORT, page 251
3 Leaves with few or no stomates, dark green to red-green or red-brown; plants aquatic, usually submerged; girdle evident (but obscure in *I . prototypus*) 4

4   Megaspores averaging more than 0.6 mm wide (ranging from 0.55–0.75 mm); leaves abruptly tapered to the tip; plants typically submerged in water 1–3 m deep; CT, MA, ME, NY, RI, VT .... *Isoetes lacustris*, LAKE QUILLWORT, page 250

4   Megaspores averaging less than 0.6 mm (ranging from 0.4–0.65 mm); leaves gradually tapered to the tip; plants submerged to about 1 m deep 5

5   Velum covering the entire sporangium; leaves very rigid and straight to the tip; sporangium wall unpigmented; girdle of the megaspore obscure; ME............ *Isoetes prototypus* .... SPIKE QUILLWORT, page 247

5   Velum covering less than half the sporangium; leaves flexible and curled at tip; sporangium wall streaked with brown; girdle present; regionwide but absent from PA .... *Isoetes tuckermanii*, TUCKERMAN'S QUILLWORT, page 251

ADDITIONAL NORTHEASTERN SPECIES

• *Isoetes appalachiana* D.F. Brunton & D.M. Britton (**Appalachian quillwort**); formerly considered part of a broadly defined *I. engelmannii;* found in Pennsylvania and New Jersey in acidic lakes and shores; velum covers up to 1/4 of the sporangium.

• *Isoetes prototypus* D.M. Britt. (**Spike quillwort**); rare, submerged in cold, clear lakes in eastern Maine (threatened); plants small, to 15 cm tall; velum completely covers sporangium.

• *Isoetes valida* (Engelm.) Clute (**True quillwort**); formerly treated as *I. engelmannii* var. *caroliniana;* known from lakes, streams, and swamps in Pennsylvania and southern New York; velum covers 1/3–2/3 of the sporangium.

• *Isoetes viridimontana* M.A. Rosenthal & W.C. Taylor (**Green Mountain quillwort**), endemic to Vermont and reported reported from Windham County.

Isoetes appalachiana
APPALACHIAN QUILLWORT

HYBRIDS

Many *Isoetes* hybrids have been reported from the region, and may be present in habitats where several species of quillwort co-occur; hybrids often have megaspores of variable size, some of which may be deformed.

Isoetes lacustris

Isoetes echinospora Durieu
SPINY-SPORE QUILLWORT

DISTRIBUTION CT | MA | ME | NH | NJ | NY | PA | RI | VT

FIELD TIPS
- region's most common quillwort
- plants usually submerged, sometimes emergent in shallow water along shores
- leaves variable

HABITAT
Shallow, somewhat acidic water and shores of lakes, ponds and streams with sandy or gravelly bottoms.

DESCRIPTION

LEAVES Bright green to reddish green, pale toward base, often deciduous, to about 30 cm long.

MEGASPORES White, covered with sharp spines; spores mature late summer.

SYNONYMS *Isoetes braunii* Durieu, *Isoetes muricata* Durieu

NOTE May hybridize with *Isoetes engelmannii* to form sterile hybrid **Isoetes × eatonii** Dodge (pro sp.). *I. echinospora* is the most widely distributed quillwort in North America.

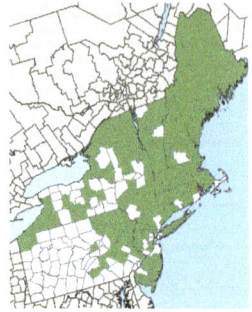

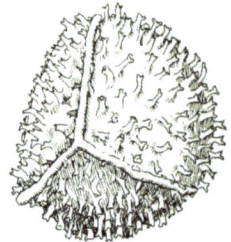

megaspore

VALERII GLAZUNOV

Isoetes engelmannii A. Braun
ENGELMANN'S QUILLWORT

DISTRIBUTION CT | MA | ME | NH | NJ | NY | PA | RI | VT

FIELD TIPS
- region's largest quillwort
- leaves to about 60 cm long, bright green, evergreen

HABITAT
Shallow water of lakes and rivers; plants submerged or emergent.

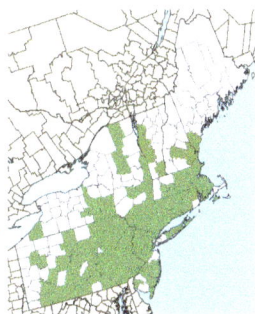

DESCRIPTION

LEAVES Bright green, pale toward base, evergreen, to about 60 cm long or sometimes longer, can be sprawling.

MEGASPORES White; surface ridged and honey-combed; spores mature in summer.

CONSERVATION STATUS *endangered* New Hampshire, *threatened* Vermont.

NAME First collected in Missouri by Engelmann, and named in his honor by A. Braun in 1846.

megaspore

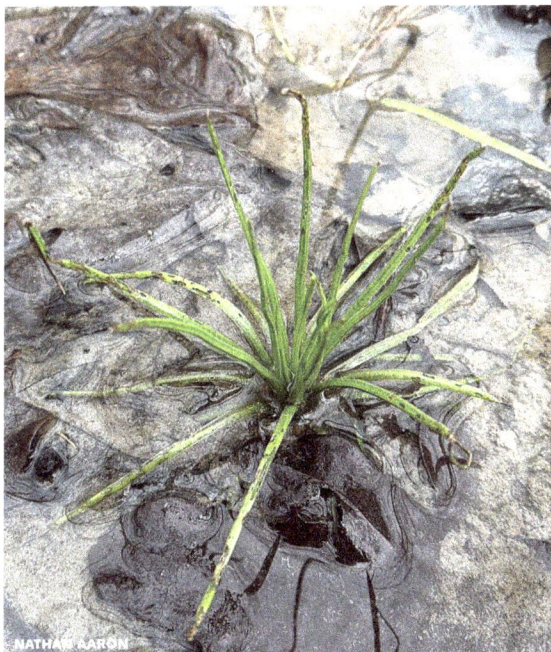

NATHAN AARON

Isoetes lacustris L.
WESTERN LAKE QUILLWORT

DISTRIBUTION CT | MA | ME | NH *historical* | NJ | NY | PA | RI | VT

FIELD TIPS
• submerged in acidic ponds and lakes
• leaves short, evergreen, stiff

HABITAT

Submerged in cold, clear, nutrient poor, acidic ponds and lakes with sandy or gravelly bottoms; found at depths up to 3 m or more. Unlike *Isoetes echinospora,* western lake quillwort almost always found in deep cold water rather than emergent on shores.

DESCRIPTION

LEAVES Dark green, pale brown near base, evergreen, stiff, abruptly tapered to tip, mostly 5–10 cm long (occasionally to 20 cm long; velum covers up to 1/2 the sporangia.

MEGASPORES White, sporadically covered with netlike ridges and with a distinct ridged midrib; spores mature late summer.

SYNONYMS *Isoetes macrospora* Durieu, *Isoetes hieroglyphica* A.A. Eaton

CONSERVATION STATUS *endangered* Massachusetts, New Hampshire, New Jersey.

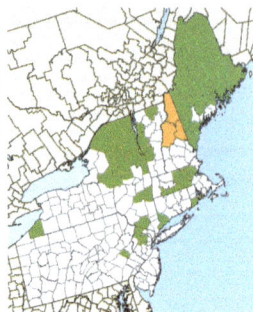

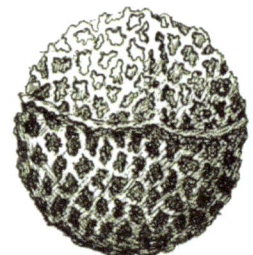

megaspore

Isoetes riparia Engelm. ex A. Braun
SHORE QUILLWORT

DISTRIBUTION CT | MA | NJ | NY | PA | RI | VT | ME, NH *historical*

FIELD TIPS
• submerged in shallow water of rivers, ponds, lakes, estuaries and tidal shores

HABITAT
Sandy, cobbly, and muddy margins of rivers and lakes, including tidal shores; water slightly acidic to circumneutral.

DESCRIPTION

LEAVES Deep green, somewhat rigid, 10–20 cm long; velum covers about 1/3 of the sporangia.

MEGASPORES White; surface spiny or crested, without a girdle.

CONSERVATION STATUS *endangered* New Hampshire, New York.

NOTE As there is some confusion in herbarium records, in this treatment, *Isoetes riparia* includes *Isoetes septentrionalis* D.F. Brunton (Northern quillwort), formerly treated as *Isoetes riparia* var. *canadensis* Engelm.

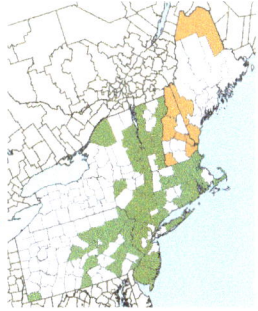

Isoetes tuckermanii A. Braun
TUCKERMAN'S QUILLWORT

DISTRIBUTION CT | MA | ME | NH | NJ | NY | RI | VT

FIELD TIPS
• pond and lake margins, shores (often tidal), usually submersed in quiet water.
• leaves olive-green, recurved at tip

HABITAT
Submerged (or leaves sometimes emergent) in slightly acidic water of lakes and slow-moving streams; often on shores and estuaries where subject to tidal fluctuations of water levels.

DESCRIPTION

LEAVES Olive-green, stout, stiff, recurved at tip.

MEGASPORES White, surface with netlike ridges and densely bumpy girdle along lower side of equatorial ridge.

SYNONYM *Isoetes acadiensis* Kott.

SIMILAR SPECIES *Isoetes lacustris,* which has larger megaspores and commonly grows in water 1 m or more deep (vs. *I. tuckermanii,* with similar megaspores but smaller (0.4-0.65 mm in diameter), and commonly growing in water shallower than 1 m deep.

megaspore
Isoetes riparia

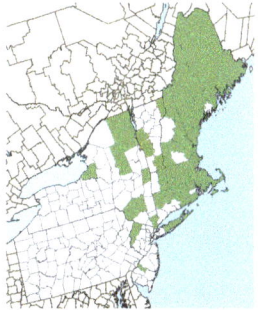

LYCOPODIACEAE *Clubmoss family*

MOSTLY EVERGREEN, PERENNIAL HERBS, somewhat mosslike (but larger) and some genera resembling miniature conifer trees. Stems upright, trailing or creeping, most species with stems lying flat on the ground and with upright shoots terminating in cylindrical, spore-producing strobili (cones, these absent in *Huperzia*); covered by many small, linear to lance-shaped, 1–veined leaves. Gametophytes small, and in all but *Lycopodiella,* are produced underground, lack chlorophyll and are dependent upon mycorrhizal fungi; gametophytes of *Lycopodiella* develop on the surface and are photosynthetic. Earlier treatments of the family recognized just 2 genera: *Lycopodium* (North America) and *Phylloglossum* (Australia and New Zealand). Now, the family is separated into 16 genera worldwide (7 genera in our flora), and an estimated 388 species.

NOTE Dried clubmoss spores have water-repellent properties, and were sold under the name lycopodium. Uses included dusting wounds and chafed skin in infants, and dusting pills to prevent their sticking to one another. Spores were also used in fireworks; when placed in fire, the spores ignite, producing a bright light.

ADDITIONAL NORTHEASTERN SPECIES
• *Pseudolycopodiella caroliniana* (L.) Holub (**Carolina false clubmoss**); wet, usually sandy soils; reported from New Jersey and New York (endangered); historically known from a single site in Massachusetts, becoming more common in southeastern USA; sporophylls of strobilus yellowish, wider and shorter than the erect stem leaves.

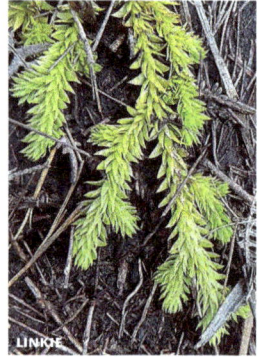

Pseudolycopodiella caroliniana

4　Aerial stems much branched and "tree-like," with the branches often spreading; strobili usually several per upright shoot........ *Dendrolycopodium*, TREE-CLUBMOSS, page 254

5　Leaves spreading, lance-shaped, tipped with a whitish, hairlike bristle ...... *Lycopodium* .... GROUND-PINE, page 278

5　Leaves appressed, partially fused with stem, ± scalelike, not tipped with a whitish bristle. .... *Diphasiastrum*, CREEPING-CEDAR, page 259

MATT BE...

Dendrolycopodium obscurum PRINCESS-PINE

Dendrolycopodium A. Haines TREE-CLUBMOSS

PLANTS BRANCHED AND RESEMBLE SMALL TREES; horizontal stems buried in soil. Branches round or compressed in cross-section. Leaves 6- to 8-ranked, tapered to a point at tip but not spine- or hair-tipped. Strobili (cones) stalkless, mostly 1–7 at ends of upright stem branches. Worldwide, 4 species in genus, 3 of which occur mostly in eastern North America and in our flora.

KEY CHARACTERS
• Plants treelike, the main axis with forked branches.
• Branches mostly round in cross-section.
• Strobili (cones) not stalked.

NOTE *Dendrolycopodium dendroideum* and *D. hickeyi* long considered to be varieties of princess-pine (*D. obscurum*) and the 3 species can be difficult to distinguish, especially in dried specimens.

NAME From Greek, *dendro,* tree, referring to treelike form of these clubmosses.

KEY TO *DENDROLYCOPODIUM* | TREE-CLUBMOSS

1 Branches somewhat flattened in cross-section; leaves on one of the flattened faces smaller than the others; leaves tightly appressed (not spreading) on the main stem below the branches; common, regionwide . . . . *Dendrolycopodium obscurum*
. . . . PRINCESS-PINE, page 258

1 Branches round in cross-section; leaves all the same size; leaves appressed or spreading on the main stem 2

2 Leaves arranged in 2 ranks on each lateral edge of a branch, with 1 rank each on upper and lower surface; leaves on lower stem tightly appressed, dark green; regionwide . . . . . .
. . . . *Dendrolycopodium hickeyi*, PENNSYLVANIA TREE-CLUBMOSS, page 257

2 Leaves arranged in 1 rank on each lateral edge of a branch, with 2 ranks on upper and lower surfaces; leaves on lower stem spreading, pale green; regionwide . . . .
. . . . *Dendrolycopodium dendroideum*, PRICKLY TREE-CLUBMOSS, page 256

SUMMARY OF CLUBMOSS GENERA

NEW GENUS NAME	FORMER NAME	COMMON NAME
Dendrolycopodium	*Lycopodium obscurum* group	Tree-clubmoss
Diphasiastrum	*L. complanatum* group	Creeping Cedar
Huperzia	*L. selago* group	Fir-moss
Lycopodiella	*L. inundatum* group	Bog Clubmoss
Lycopodium	*L. clavatum* group	Ground-pine
Spinulum	*L. annotinum* group	Bristly Clubmoss

MATT BERGER

Dendrolycopodium dendroideum PRICKLY TREE-CLUBMOSS

Dendrolycopodium dendroideum (Michx.) A. Haines
PRICKLY TREE-CLUBMOSS

DISTRIBUTION CT | MA | ME | NH | NJ | NY | PA | RI | VT

FIELD TIPS
- evergreen, treelike clubmoss
- leaves spreading, stiff and prickly
- strobili (cones) stalkless, single at branch ends

HABITAT

Moist to dry deciduous and mixed forests; may form large colonies; soils sandy and acidic.

DESCRIPTION

STEMS Spreading by underground horizontal stems, with branched, upright shoots to ca. 30 cm tall, appearing somewhat like a small tree; branches round in cross-section.

LEAVES Spreading to ascending, linear; leaves of lower stems (below the branches) stiff and needlelike (prickly to touch); leaves of branches 6-ranked.

STROBILI Stalkless, single at branch ends, with 1–7 or more on each upright stem.

SYNONYMS *Lycopodium dendroideum* Michx., *Lycopodium obscurum* var. *dendroideum* (Michx.) D.C. Eaton, *Lycopodium obscurum* var. *hybridum* Farw.

SIMILAR SPECIES The spreading, prickly leaves on lower stems distinguish prickly tree-clubmoss from our other 2 species. Also, the lateral branches are nearly round, and not conspicuously flattened as in princess-pine (*D. obscurum*).

NAME When Michaux found this on his travels from "Canada to the Carolina mountains"' he considered it a new species and in 1803 named it *Lycopodium dendroideum*. It is so close to the one Linnaeus had previously named *L. obscurum*, that in 1890 D. C. Eaton made Michaux's plant a variety. It is currently treated, once again, as a valid species.

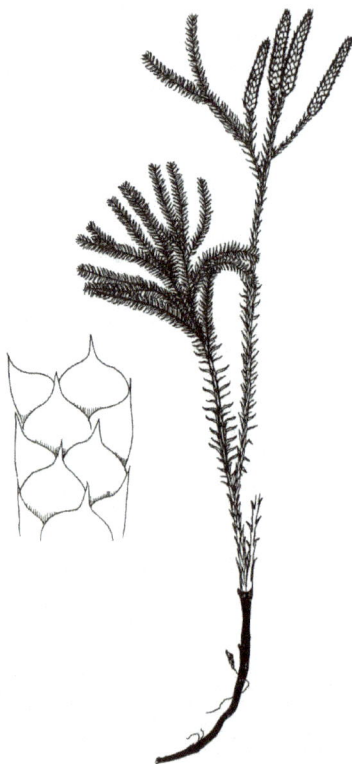

Dendrolycopodium hickeyi (W.H. Wagner, Beitel & Moran) A. Haines
PENNSYLVANIA TREE-CLUBMOSS

DISTRIBUTION CT | MA | ME | NH | NJ | NY | PA | RI | VT

FIELD TIPS
- evergreen, treelike clubmoss
- base of upright stem not prickly
- leaves spreading and ascending
- strobili (cones) stalkless, 1–7 per each upright stem

HABITAT

Drier deciduous, mixed, and conifer forests; may form colonies; soils sandy and acidic.

DESCRIPTION

STEMS Spreading by underground horizontal stems, with branched, upright shoots to ca. 20 cm tall, appearing somewhat like a small tree; branches round to elliptic in cross-section.

LEAVES All of about equal size, widest near middle, tapered to a pointed tip; margins entire; branch leaves 6-ranked, spreading; stem leaves below branches appressed.

STROBILI Stalkless, 1–7 on each upright stem.

SYNONYMS *Lycopodium hickeyi* W.H. Wagner, Beitel & Moran, *Lycopodium obscurum* var. *isophyllum* Hickey

SIMILAR SPECIES **Princess-pine** (*D. obscurum*) has branches flattened in cross-section, with twisted branch leaves of different sizes; in *D. hickeyi,* branches are nearly round, and the leaves are all similar size and not twisted.

NOTE *D. hickeyi* is was first recognized in 1977 as *Lycopodium obscurum* var. *isophyllum,* and now usually treated as a valid species.

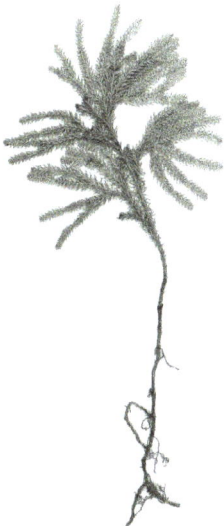

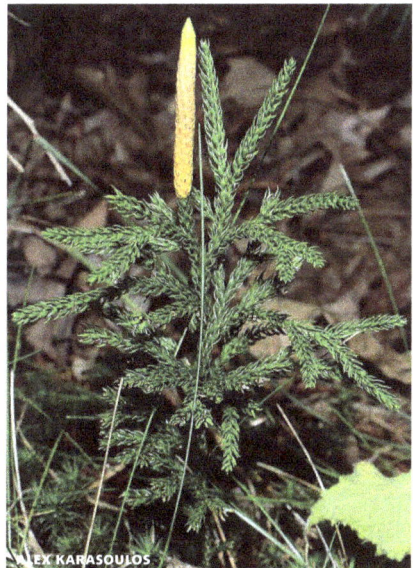

ALEX KARASOULOS

Dendrolycopodium obscurum (L.) A. Haines
PRINCESS-PINE

DISTRIBUTION CT | MA | ME | NH | NJ | NY | PA | RI | VT

FIELD TIPS
- evergreen, treelike clubmoss
- main stem not prickly
- leaves ascending, 6-ranked
- strobili (cones) stalkless, single at branch ends,
  1–6 per each upright stem

HABITAT

Moist to dry deciduous and mixed conifer-deciduous forests, and forest margins; may form large colonies; soils sandy and acidic.

DESCRIPTION

STEMS Spreading by underground horizontal stems, with branched, upright shoots to ca. 30 cm tall, appearing somewhat like a small tree; branches flattened in cross-section or if viewed from the side.

LEAVES Ascending, narrowly lance-shaped, tapered to a pointed tip; margins entire; branch leaves 6- to 8-ranked, lateral leaves twisted, leaves along lower surface of branch smaller than the others; stem leaves below branches appressed.

STROBILI Stalkless, 1–6 on each upright, branched stem.

SYNONYM *Lycopodium obscurum* L.

SIMILAR SPECIES
- **Prickly tree-clubmoss** (*D. dendroideum*) has lower stems (below the branches) prickly to the touch.
- **Pennsylvania tree-clubmoss** (*D. hickeyi*) has branches nearly round in cross-section, and branch leaves all of similar size.

NAME Collected near Philadelphia by John Bartram, and sent to Europe, where a drawing of it was published by Dillenius in 1741. Linnaeus named it *Lycopodium obscurum* in 1753.

TOM SCAVO

Diphasiastrum Holub CREEPING-CEDAR

SMALL PLANTS OF DRIER HABITATS resembling miniature trees; branches flattened or 4-angled in cross-section (round in *D. sitchense*); leaves 4-ranked (5-ranked in *D. sitchense*), neither spine- nor hair-tipped. Strobili (cones) stalked, the stalks branched into segments of equal length. Worldwide, 20 species defined; five species in North America, four species (and several hybrids) in our flora.

KEY CHARACTERS

- Plants treelike, the main axis with forked branches.
- Branches flattened or 4-angled in cross-section.
- Leaves keeled and also unkeeled along midrib.
- Strobili (cones) on long stalks.

NAME *Diphasium* is another genus in Clubmoss Family; *astrum* refers to a 'partial resemblance.'

HYBRIDS

- *Diphasiastrum* × *habereri* (House) Holub, hybrid between *D. digitatum* and *D. tristachyum;* regionwide.
- *Diphasiastrum* × *sabinifolium* A.V. Gilman, hybrid between *D. sitchense* and *D. tristachyum;* see description, page 259.
- *Diphasiastrum* × *verecundum* A.V. Gilman, hybrid between *D. complanatum* and *D. digitatum;* uncommon in region, absent from Connecticut, New Jersey, and Rhode Island..
- *Diphasiastrum* × *zeilleri* (Rouy) Holub, hybrid between *D. complanatum* and *D. tristachyum;* northern portions of region.

BONNIE SEMMLIN

Diphasiastrum × *habereri*

KEY TO *DIPHASIASTRUM* | CREEPING-CEDAR

1 Leaves of the lateral branches all alike, arranged in 5 ranks, with stomates on both surfaces; strobilus stalks absent (rarely to 1 cm tall); strobili solitary; plants typically less than 12 cm tall, without an evident main axis; rare in ME, NH, NY . . . . *Diphasiastrum sitchense*
. . . . SITKA CREEPING-CEDAR, page 265

1 Leaves of the lateral branches not all alike, arranged in 4 ranks, with stomates on the lower surface only; strobilus stalks 0.5–15 cm long; strobili 1–4 per stalk; plants typically more than 12 cm tall, usually dendroid (tree-like) 2

2 Strobili 1 (–2) on short to elongated, rarely forked stalks (peduncles); base of strobilus with a few sporophylls scattered along peduncle; stomata on both leaf surfaces; ME, NH, NY, VT . . . . *Diphasiastrum* × *sabinifolium*, SAVIN-LEAF CREEPING-CEDAR, page 264

2 Strobili 2–4 on forked stalks; base of strobilus compact and abruptly distinct from peduncle; stomata on lower leaf surfaces only 3

3 Horizontal stems almost always buried in soil (4 cm or more deep); ultimate (outermost) branches squarish in cross section, 1–2 mm wide; plants usually blue-green (except in deep shade); lower leaves same size as upper; regionwide . . . . *Diphasiastrum tristachyum*
. . . . DEEP-ROOT CREEPING-CEDAR, page 266

3 Horizontal stems on ground surface, or slightly buried (less than 4 cm deep) in surface litter; ultimate (outermost) branches flat in cross section; 2–4 mm wide; plants not blue-green; lower leaves reduced to a leaf base and a short free appendage 4

4 Ultimate branches symmetrically in one plane forming a layered appearance; annual constrictions mostly absent; stalks of strobili staying green until after spore discharge in autumn; common, regionwide.. *Diphasiastrum digitatum*, FAN CREEPING-CEDAR, page 262

4 Ultimate branches not evenly layered, giving an irregular appearance; annual constrictions uniformly present; stalks of strobili losing green color by time of spore discharge in late summer; ME, NH, NY, VT . . . . *Diphasiastrum complanatum*
. . . . NORTHERN GROUND-CEDAR, page 260

Diphasiastrum complanatum (L.) Holub
NORTHERN CREEPING-CEDAR

DISTRIBUTION ME | NH | NY | VT

FIELD TIPS
- plants creeping, resembling a small cedar tree
- branches flattened, bright green
- leaves 4-ranked, appressed, of different sizes
- strobili on slender stalks, these sometimes branched

HABITAT
Woods, thickets, clearings; soils typically dry, sandy and acidic.

DESCRIPTION

STEMS Horizontal stems on ground surface or shallowly underground. Upright stems to about 30–40 cm long, with crowded, irregularly forking branchlets, these flattened, and often strongly constricted between annual growths. Upper surface of branches green and satiny, underside pale and dull.

LEAVES 4-ranked, appressed and attached to stem for more than half their length; upper rank of leaves small and very narrow; lateral leaves larger, keeled; underside leaves smallest.

STROBILI Mostly 1 or 2 on slender, sometimes forked stalks; the strobili fertile for their entire length.

SYNONYMS *Diphasium anceps* (Wallr.) Á. & D. Löve, *Diphasium complanatum* (L.) Rothm., *Diphasium wallrothii* H.P. Fuchs, *Lycopodium anceps* Wallr., *Lycopodium complanatum* L.

CONSERVATION STATUS *endangered* New York.

SIMILAR SPECIES **Fan creeping-cedar** (*D. digitatum*) lacks annual constrictions on stems and branches, and its strobili are often sterile in their upper portion.

NAME Linnaeus described *Lycopodium complanatum* in 1753. From Latin, *complanatus,* made level or flat, referring to the flattened leafy branches, similar to those of a cedar tree.

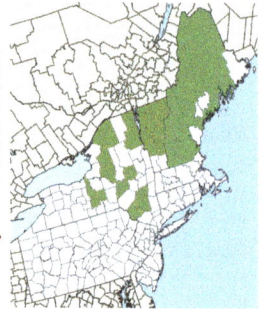

CHLOE & TREVOR VAN LOON · *Diphasiastrum complanatum* NORTHERN CREEPING-CEDAR

Diphasiastrum digitatum (Dill. ex A. Braun) Holub
FAN CREEPING-CEDAR

DISTRIBUTION CT | MA | ME | NH | NJ | NY | PA | RI | VT

FIELD TIPS
- creeping, resembling a small cedar tree
- branches flattened, dark green
- leaves 4-ranked, of different sizes, appressed or spreading at tips
- strobili stalked, the stalks usually twice-forked

HABITAT

Sandy woods, clearings, fields, thickets, where sometimes forming extensive patches from the shallow rhizomes; soils typically sandy and acidic.

DESCRIPTION

STEMS Horizontal stems mostly on ground surface or in surface litter. Upright stems 30–40 cm long, with fanlike branches, constrictions between annual growth absent or indistinct; upper surface of branches green and shiny, underside pale and dull.

LEAVES Tiny, 4-ranked, appressed and attached to stem for more than half their length; upper rank of leaves small and very narrow; lateral leaves larger, keeled, spreading at tip; underside leaves smallest, triangular.

STROBILI 2–4, on stalks usually twice-forked from near the same point; tips of the strobili are often sterile (check a number of plants to verify this).

SYNONYMS *Diphasium complanatum* subsp. *flabelliforme* (Fernald) Á. & D. Löve, *Lycopodium complanatum* var. *flabelliforme* Fernald, *Lycopodium digitatum* Dill. ex A. Braun, *Lycopodium flabelliforme* (Fernald) Blanch.

SIMILAR SPECIES Northern creeping-cedar (*D. complanatum*) has distinct annual constrictions on stems and branches, and the strobili are completely fertile.

NOTE Another of the Philadelphia plants collected by John Bartram and illustrated by Dillenius in 1741. For many years thereafter it was confused with the more northern *Diphasiastrum complanatum,* but Fernald segregated it as a variety in 1909, and Blanchard raised it to species rank in 1919.

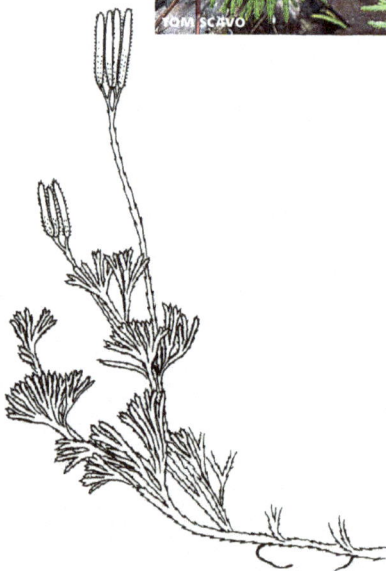

MATT BERGER

Diphasiastrum digitatum FAN CREEPING-CEDAR

Diphasiastrum × sabinifolium (Willd.) Holub
SAVIN-LEAF CREEPING-CEDAR

DISTRIBUTION ME | NH | NY PA *historical* | VT

FIELD TIPS
- plants creeping, resembling a small cedar tree
- branches compressed in cross-section
- leaves 4-ranked, of similar size
- strobili on slender stalks, these sometimes branched
- sporophylls on lower portion of strobili widely spaced

HABITAT

Sandy woods and meadows, often where disturbed (as in blowout areas).

DESCRIPTION

STEMS Horizontal stems mostly creeping on ground surface. Erect stems dichotomously forked, up to 20 cm long; sterile branchlets flattened.

LEAVES 4-ranked, awl-like, nearly similar lengths; leaves of upper and lower side appressed; lateral leaves slightly larger and appressed to stem for about half their length, the free portion spreading and incurved at tip.

STROBILI Stalked, the stalks sometimes forked; sporophylls near base of strobili widely spaced.

SYNONYMS *Lycopodium armatum* Desv., *Lycopodium sabinifolium* Willd.

NOTE Considered to have originated as a hybrid between *Diphasiastrum sitchense* and *D. tristachyum. Diphasiastrum sitchense* differs from *D. × sabinifolium* in having sessile rather than stalked strobili.

NAME Sabina is a section of the conifer genus *Juniperus* (commonly called 'savin'); the leaves of this species are scalelike and similar to those of juniper.

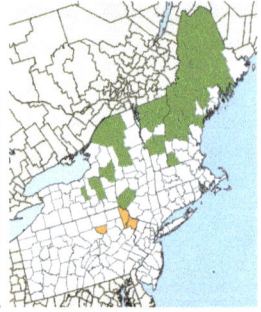

CHRISTIAN GRENIER

NATE MARTINEAU

Diphasiastrum sitchense (Rupr.) Holub
SITKA CREEPING-CEDAR

DISTRIBUTION ME | NH | NY

FIELD TIPS
- plants creeping, resembling a small cedar tree
- sterile branches round in cross-section
- leaves 5-ranked, of similar size
- strobili sessile
- mostly found at higher elevations

HABITAT

Moist, rocky openings, alpine and subalpine barrens and thickets; open conifer forests and grassy clearings.

DESCRIPTION

STEMS Horizontal stems on substrate surface or shallowly buried, with distant, yellowish, scale-like leaves; upright stems clustered and branching near base, mostly 5–15 cm tall; sterile branchlets dark green, somewhat shiny, round in cross-section; annual bud constrictions inconspicuous.

LEAVES Appressed, lance-shaped, 3–5 mm long, 5-ranked, not overlapping, adnate for less than 1/2 their length, the free part usually incurved at tip.

STROBILI Sessile on leafy branches, not on naked peduncles (or rarely on short peduncles to 1 cm long).

SYNONYMS *Lycopodium armatum* Desv., *Lycopodium sabinifolium* Willd., *Lycopodium sitchense* Rupr.

CONSERVATION STATUS *endangered* New Hampshire, New York.

SIMILAR SPECIES *Diphasiastrum sitchense* differs from *D. × sabinifolium* in having sessile rather than stalked strobili, and round rather than flattened sterile branches.

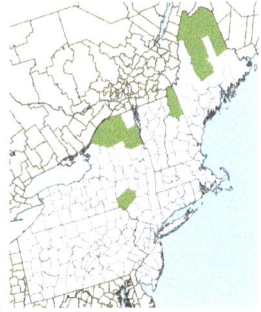

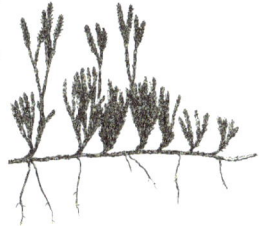

Diphasiastrum sitchense SITKA CREEPING-CEDAR

Diphasiastrum tristachyum (Pursh) Holub
DEEP-ROOT CREEPING-CEDAR

DISTRIBUTION CT | MA | ME | NH | NJ | NY | PA | RI | VT

FIELD TIPS
- plants blue-green, in crowded clusters
- branches somewhat 4-angled, fanlike, with annual constrictions
- leaves 4-ranked, appressed, of different sizes
- strobili on slender stalks, these once- or twice-forked, rising above leafy branches

HABITAT
Dry sandy fields and open woods; sandy disturbed places.

DESCRIPTION

STEMS Horizontal stems usually deeply buried. Upright stems to about 30 cm long. Sterile branches upright to loosely spreading, somewhat 4-angled in cross-section, with small annual constrictions.

LEAVES Leaves tiny, narrow, 4-ranked, evergreen, bluish green and usually covered with a fine white powder, attached to stem for more than half their length; upper and lower rank of leaves about half the size of lateral leaves.

STROBILI Mostly 3–4 on leafy-bracted stalks, the stalks once- or twice-forked.

SYNONYMS *Diphasium chamicyparissus* (A. Braun) Á. & D. Löve, *Diphasium complanatum* subsp. *chamicyparissus* (A. Braun) Kukkonen, *Diphasium tristachyum* (Pursh) Rothm., *Lycopodium chamicyparissus* A. Braun, *Lycopodium complanatum* subsp. *chamicyparissus* (A. Braun) Nyman, *Lycopodium tristachyum* Pursh

SIMILAR SPECIES The blue-green color, squarish stems, and conspicuous constrictions at the juncture between each year's growth help separate deep-root creeping-cedar from other members of the genus. Note that plants growing in shade tend to be more green in color than blue-green, and more openly branched than those found in open places.

NAME This species grows both in Europe and eastern North America, but was first named from plants collected in Virginia, by Pursh in 1814. Later named in Europe *Lycopodium chamaecyparissus.*

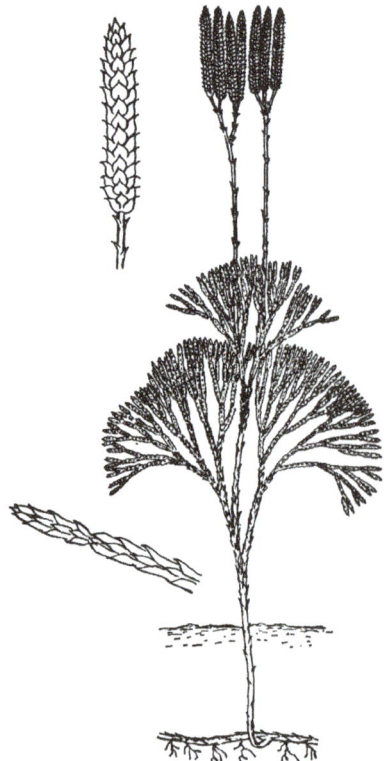

JESSICA UTRUP

Diphasiastrum tristachyum DEEP-ROOT CREEPING-CEDAR

Huperzia Bernh. **FIR-MOSS**

EVERGREEN PLANTS WITH SMALL, STALKLESS, POINTED LEAVES; horizontal stems absent. Sporangia borne at bases of stem leaves, rather than in terminal strobili (cones) as in the other members of Family. *Huperzia* also produce gemmae (small plantlets) in upper portion of stem; if the gemmae land on a favorable site they can generate roots and develop into a new plant, identical to the parent plant. Genus contains 25 species worldwide; seven in North America, four in our flora.

KEY CHARACTERS
- Plants with bristly upright stems with sporangia borne in distinct zones; strobili (cones) absent.
- Horizontal stems absent (but older stems sometimes lean and appear horizontal).
- Gemmae (plantlets) produced in upper parts of mature stems.

HYBRIDS
- *Huperzia* × *bartleyi* (Cusick) Kartesz & Gandhi, hybrid between *H. lucidula* and *H. porophila;* reported from Ulster County, New York.
- *Huperzia* × *buttersii* (Abbe) Kartesz & Gandhi, hybrid between *H. lucidula* and *H. selago;* known from Massachusetts, Maine, and Vermont; resembles a slender *H. lucidula,* but distinguished by presence of abortive spores, and scattered stomata on upper leaf surface.
- *Huperzia* × *josephbeitelii* A. Haines, hybrid between *H. appalachiana* and *H. selago;* Maine, New Hampshire, Vermont.
- *Huperzia* × *protoporophila* A. Haines, hybrid between *H. appressa* and *H. lucidula;* Massachusetts, Maine, New Hampshire, Vermont.

KEY TO *HUPERZIA* | FIR-MOSS

1   Leaves obovate, widest above the middle, spreading to ± reflexed, the upper portion of at least the larger leaves with distinct teeth; shoots "shaggy" with conspicuous annual constrictions; usually growing on soil; regionwide . . . . *Huperzia lucidula*
. . . . SHINING FIR-MOSS, page 270

1   Leaves lance-shaped, widest below the middle, leaves (at least those on the upper stem) often ascending, entire or with a few small teeth; annual constrictions absent or faint . . 2

2   Leaves lance-shaped with sides nearly parallel; stomates on upper surface of each leaf number 2–50 (view fresh leaves under 20x lens to see the light-colored, dot-like stomates); rare in eastern PA . . . . *Huperzia porophila*, ROCK FIR-MOSS, page 272

2   Leaves lance-shaped (as above) or ovate or triangular; if leaf shape is inconclusive, then number of stomates on upper leaf surface greater than 60; more widespread in region . 3

3   Leaves near base of plant essentially same size as those on upper portion; gemmae formed in a single whorl at end of the annual growth; MA, ME, NH, VT . . . . *Huperzia selago*
. . . . FIR-MOSS, page 273

3   Leaves near base of plant conspicuously longer than those on upper portion; gemmae formed throughout upper portions of shoot; CT, MA, ME, NH, NY, VT *Huperzia appressa*
. . . . MOUNTAIN FIR-MOSS, page 269

Huperzia appressa (Desv.) Á. & D. Löve
MOUNTAIN FIR-MOSS

DISTRIBUTION CT | MA | ME | NH | NY | VT

FIELD TIPS
- short, evergreen plants of rocky places
- leaves yellow-green to green, reduced in size on upper portion of stem
- sporangia and gemmae borne on upper stems

HABITAT
Cliffs, talus slopes, where open and exposed, on moss or thin soil.

DESCRIPTION

STEMS Short, to only 10 cm long; clustered; annual constrictions absent.

LEAVES Ascending, narrowly lance-shaped or with the sides parallel; stomates on both surfaces; margins entire; upper stem leaves smaller than those of lower stem.

SPORANGIA In distinct zones on upper stems.

SYNONYM *Huperzia appalachiana* Beitel & Mickel

CONSERVATION STATUS *endangered* Massachusetts, New Hampshire.

SIMILAR SPECIES **Fir-moss** (*Huperzia selago*) produces gemmae in a single whorl near end of each year's growth; gemmae in mountain fir-moss are scattered throughout much of stem.

Huperzia appressa MOUNTAIN FIR-MOSS

Huperzia lucidula (Michx.) Trevisan
SHINING FIR-MOSS

DISTRIBUTION CT | MA | ME | NH | NJ | NY | PA | RI | VT

FIELD TIPS
- region's most common and widespread fir-moss
- leaves evergreen, shiny, usually widest above middle, margins toothed
- stems with distinct annual constrictions

HABITAT

Moist deciduous, mixed conifer-hardwood, or conifer forests, often sprawling on mossy boulders, logs, and hummocks; less commonly on shaded, mossy sandstone.

DESCRIPTION

STEMS Ascending and sprawling, single or few-forked, to ca. 20 cm long, rooting towards base; annual constrictions evident; lower stem often hidden in leaf litter; sometimes forming a circular "fairy ring" as the stems grow outward.

LEAVES Mostly 6-ranked, 7–12 mm long, oblong lance-shaped, widest above middle, spreading or angled downward; stomates only visible on underside; margins toothed, especially above middle of leaf.

SPORANGIA In upper portion of stem, yellow when mature; gemmae in single whorl from uppermost leaf axils.

SYNONYMS *Huperzia selago* subsp. *lucidula* (Michx.) Á. & D. Löve, *Lycopodium lucidulum* Michx., *Lycopodium reflexum* Sw., non Lam., *Urostachys lucidulus* (Michx.) Herter ex Nessel

NOTE Shining fir-moss is a distinctive and usually readily identified species. Key characters include toothed leaves widest above the middle, stems with conspicuous annual constrictions, and stomata only on lower surface of the leaves (visible under magnification). *Huperzia lucidula* is our most common and widespread fir-moss, usually found in forest soil rather than on rocks.

NAME In the course of his travels, Michaux observed this plant from "Canada to the Carolina mountains," and named it *Lycopodium lucidulum* in 1803.

ÉTIENNE LÉVEILLÉ-BOURRET

Huperzia lucidula SHINING FIR-MOSS

Huperzia porophila (Lloyd & Underwood) Holub
ROCK FIR-MOSS

DISTRIBUTION PA

FIELD TIPS
- shaded sandstone habitats
- stems seldom branched
- leaves narrow, mostly entire, not shiny

HABITAT

Sandstone outcrops and boulders and rocky woods, where cool, moist and shaded.

DESCRIPTION

STEMS Clustered, to 15 cm long; annual constrictions faint or absent.

LEAVES Narrow, ca. 3–8 mm long; margins nearly entire.

SPORANGIA Borne from upper leaf axils.

SYNONYMS *Huperzia selago* var. *patens* (P. Beauv.) Trevis., *Huperzia selago* var. *porophila* (Lloyd & Underw.) Á. & D. Löve, *Lycopodium lucidulum* Michx. var. *porophilum* (Lloyd & Underw.) Clute, *Lycopodium porophilum* Lloyd & Underw., *Lycopodium selago* subsp. *patens* (P.Beauv.) Calder & Roy L. Taylor, *Lycopodium selago* var. *porophilum* (Lloyd & Underw.) Clute

CONSERVATION STATUS *endangered* Pennsylvania.

SIMILAR SPECIES Rock fir-moss characterized by leaves with nearly parallel sides and stomates on upper leaf surface numbering fewer than 50 (requires microscope). Often confused with **shining fir-moss** (*Huperzia lucidula*), but rock fir-moss rare in the region, plants usually smaller and more compact, the leaves lance-shaped and not shiny, with stomates on upper surface, and the ascending stems seldom branched.

NOTE This species is of hybrid origin, the parents believed to be *Huperzia lucidula* and *H. selago*.

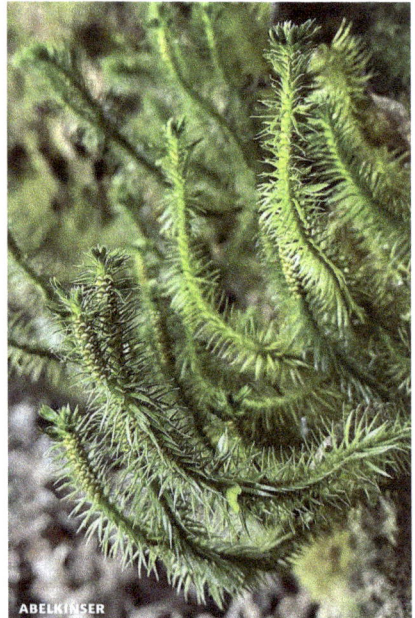

ABELKINSER

Huperzia selago (L.) Bernh. ex Mart. & Schrank
FIR-MOSS

DISTRIBUTION MA | ME | NH *historical* | NY | VT

FIELD TIPS
• plants in dense, flat-topped clusters
• stems lack annual constrictions
• gemmae at stem tip
• leaves small, entire, sharp-pointed

HABITAT
Conifer swamps, moist to dry conifer woods; mossy, shaded streambanks, boulders, trails and old roads, moist sandy borrow pits.

DESCRIPTION

STEMS Horizontal stems short, leafy; erect stems to 15 cm long, branched several times, usually near the base; annual constrictions barely evident.

LEAVES Crowded, 8- to 10-ranked, ascending, yellow-green, 3–8 mm long, lance-shaped, with stomates on both surfaces; margins entire or nearly so.

SPORANGIA In zones in upper stems; gemmae (reproductive buds) in a single whorl in upper leaf axils, and tip of stem can appear thicker than rest of stem.

SYNONYM *Lycopodium selago* L.

CONSERVATION STATUS *endangered* Massachusetts, New Hampshire, New York (historical records only); *threatened* Maine.

SIMILAR SPECIES Distinguished from shining fir-moss (*Huperzia lucidula*) by the entire leaves (rather than leaves toothed at tip). Distinguished from mountain fir-moss (*Huperzia appressa*) by the longer leaves, and the gemmae restricted to a single whorl near the end of each years growth.

NAME This plant was well-known to Linnaeus and was assigned genus and species name of *Lycopodium selago* in his Species Plantarum in 1753; 'selago' referring to South African plants of genus *Selago,* the plants heather-like.

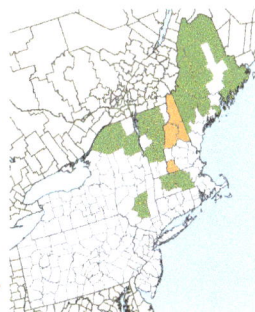

Lycopodiella Holub **BOG CLUBMOSS**

SMALL PLANTS OF WET PLACES WITH HORIZONTAL STEMS creeping on ground surface, and fertile, upright stems. Strobili (cones) formed on upper part of upright stems, stalkless and covered with leaves. Gametophytes grow on soil surface and are photosynthetic. Worldwide, 15 species in genus; 6 species in North America, 4 species in our flora, with northern bog clubmoss (*Lycopodiella inundata*) the most common.

KEY CHARACTERS
• Plants small, with slender upright stems tipped by strobili (cones).
• Horizontal stems not flattened.
• Leaves of horizontal stem all alike.

ADDITIONAL NORTHEASTERN SPECIES Prostrate clubmoss
(*Lycopodiella margueritae* J.G. Bruce, W.H. Wagner & Bietel); Pennsylvania (state endangered). Distinguished from *Lycopodiella appressa* and *L. subappressa* by the thicker and large strobili, and the more spreading leaves on the strobili and upright shoots; separated from *L. inundata* by its taller size, thicker stems, and toothed leaves.

Lycopodiella margueritae
PROSTRATE CLUBMOSS

HYBRIDS
• *Lycopodiella* × *copelandii* (Eig.) Cranfill, hybrid between *L. alopecuroides* and *L. appressa;* known from Connecticut, Massachusetts, New Jersey, southern New York, and Pennsylvania.
• *Lycopodiella* × *gilmanii* A. Haines, hybrid between *L. appressa* and *L. inundata;* sporadic regionwide but absent from Pennsylvania.
• *Lycopodiella* × *robusta* (R.J. Eat.) A. Haines, hybrid between *L. alopecuroides* and *L. inundata;* Maine, Massachusetts, and Pennsylvania.

KEY TO *LYCOPODIELLA* | CLUBMOSS

1   Horizontal stems less than 1 mm in diameter except at root nodes; horizontal stem leaves usually less than 6 mm long, teeth or bristles absent; erect shoots mostly less than 10 cm tall; regionwide . . . . *Lycopodiella inundata*, NORTHERN BOG CLUBMOSS, page 277

1   Horizontal stems 1–3 mm in diameter; horizontal stem leaves 4–13 mm long, margins toothed or bristle-tipped; tallest erect shoots often more than 10 cm tall 2

2   Horizontal shoots prominently arching over the substrate; sporophylls 5–9 mm long, spreading at maturity, with 1–3 slender teeth on each margin; uncommon in CT, MA, ME, NJ, NY, PA, RI . . . . *Lycopodiella alopecuroides*, FOX-TAIL CLUBMOSS, page 275

2   Horizontal shoots prostrate; sporophylls 3–5 mm long, appressed, entire or rarely with a low tooth on one margin; CT, MA, NH, NJ, NY, PA, RI . . . . *Lycopodiella appressa*
. . . . SOUTHERN APPRESSED CLUBMOSS, page 276

Lycopodiella alopecuroides (L.) Cranfill
FOX-TAIL CLUBMOSS

DISTRIBUTION CT | MA | ME | NJ | NY | PA | RI

FIELD TIPS
- horizontal stems commonly arching and rooting at their tip
- fertile stems erect, tipped by a 'bushy' strobilus
- wet, sandy habitats

HABITAT

Moist to wet sandy depressions pits, bog mats, sandy borrow pits.

DESCRIPTION

HORIZONTAL STEMS Arching, rooting at tip, to ca. 40 cm long.

LEAVES Appressed to somewhat spreading; margins sharply toothed, with 1–7 teeth on each side.

STROBILI Single, stalkless, 2–5 cm long, wider than stem and appearing like a bushy tail; sporophylls awl-like, longer than the sterile leaves below, widely spreading, lower portion toothed.

SYNONYMS *Lepidotis alopecuroides* (L.) Rothm., *Lycopodium alopecuroides* L.

CONSERVATION STATUS *endangered* Connecticut, Maine, Massachusetts, Pennsylvania, Rhode Island.

Lycopodiella alopecuroides FOX-TAIL CLUBMOSS

Lycopodiella appressa (Chapman) Cranfill
SOUTHERN APPRESSED CLUBMOSS

DISTRIBUTION CT | MA | NH | NJ | NY | PA | RI | ME, VT *historical*

FIELD TIPS
- horizontal stems creeping on soil surface
- fertile stems erect, more numerous than in the region's other species of *Lycopodiella*
- strobili narrow
- wet, often sandy habitats

HABITAT

Exposed shores of ponds, abandoned borrow pits, ditches; soils wet, sandy with some organic matter.

DESCRIPTION

STEMS Horizontal stems creeping on soil surface; upright stems 10–40 cm tall, deciduous.

LEAVES Narrow and appressed; margins finely toothed, with up to 7 teeth per side; rarely entire.

STROBILI Barely wider than sterile portion of upright stem; sporophylls appressed, mostly entire.

SYNONYMS *Lycopodium inundatum* var. *appressum* Chapm., *Lycopodium inundatum* var. *bigelovii* Tuckerm.

CONSERVATION STATUS *endangered* New Hampshire (historical records only); *threatened* Pennsylvania.

SIMILAR SPECIES Differs from *Lycopodiella inundata* by its taller fertile stems (to about 35 cm long), its tightly appressed sporophylls, and its mostly toothed leaves.

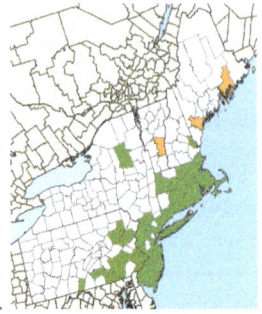

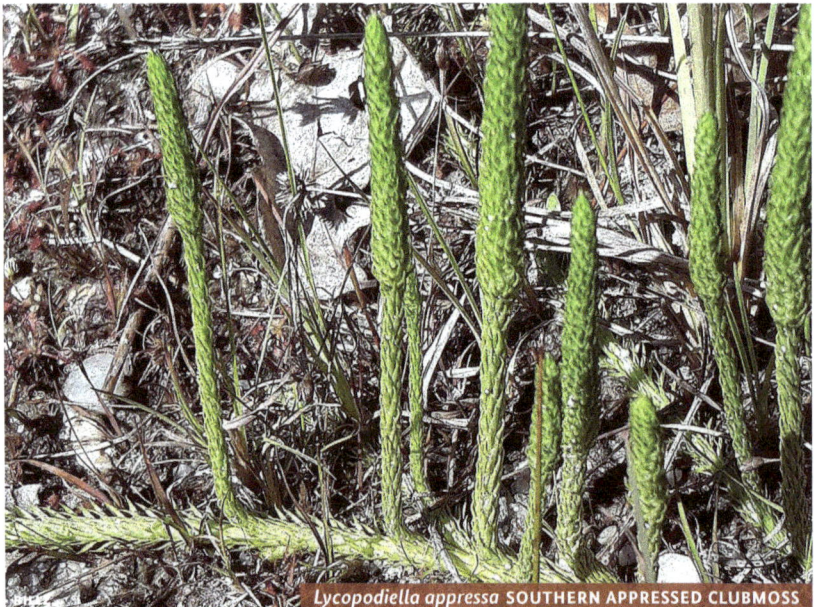

Lycopodiella appressa SOUTHERN APPRESSED CLUBMOSS

Lycopodiella inundata (L.) Holub
NORTHERN BOG CLUBMOSS

DISTRIBUTION CT | MA | ME | NH | NJ | NY | PA | RI | VT

FIELD TIPS
- horizontal stems evergreen, creeping on surface
- fertile stems erect, tipped by a 'bushy' strobilus
- wet, often sandy habitats

HABITAT

Moist to wet sandy depressions and borrow pits, bog mats, sandy lake and pond shores.

DESCRIPTION

STEMS Horizontal or arching, forking, very slender, leafy, the leaves turned upwards, rooting at intervals; fertile stems upright, to 10 cm long.

LEAVES Spiralled in 8 or 10 ranks, narrow and linear, gradually tapered to long tip; margins mostly entire.

STROBILI Single, stalkless, wider than stem and appearing like a bushy tail.

SYNONYMS *Lepidotis inundata* (L.) C. Borner, *Lycopodium inundatum* L.

NAME Being well-known in northern Europe, this plant was named *Lycopodium inundatum* by Linnaeus in 1753.

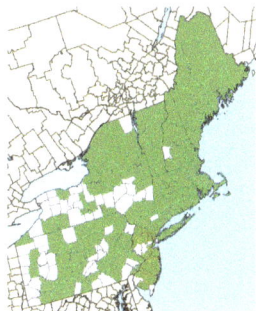

Lycopodiella inundata NORTHERN BOG CLUBMOSS

ANDREAS BERGER

Lycopodium L. GROUND-PINE

TRAILING, EVERGREEN CLUBMOSSES with clusters of upright leafy stems. Horizontal stems on or near ground surface. Upright stems branched, but not treelike as in some other members of Clubmoss Family. Leaves tipped by a long, hairlike bristle. Sporangia borne in 1 or several stalked cones at ends of stems. Gametophytes underground, disk-shaped, and nonphotosynthetic. Worldwide, 15 species are recognized (many from Asia); 2 species in North America and in our flora.

KEY CHARACTERS
- Stems upright, bristly, with several branches.
- Branches round in cross-section.
- Leaves tipped with a long, translucent hair.
- Strobili (cones) stalked, single or several in a branched cluster.

NAME From Greek, _lykos,_ a wolf, and _pous,_ a foot; perhaps the creeping stems and unequally forked branches suggestive of wolf tracks. Another common name is 'staghorn clubmoss' due to resemblance of upright branches to a stag's antlers.

KEY TO _LYCOPODIUM_ | GROUND-PINE

1   Peduncles mostly with 2–3 (–4) clearly stalked strobili; common, regionwide . . . .
    . . . . _Lycopodium clavatum_, RUNNING GROUND-PINE, page 278
1   Peduncles with 1–2 strobili, if paired, then essentially stalkless; regionwide, but absent from NJ. . . . . _Lycopodium lagopus_, ONE-CONE GROUND-PINE, page 280

Lycopodium clavatum L.
RUNNING GROUND-PINE

DISTRIBUTION CT | MA | ME | NH | NJ | NY | PA | RI | VT
FIELD TIPS
- trailing, evergreen clubmoss
- stems densely covered with small ascending leaves
- strobili (cones) usually several at ends of long, slender, branched stalk

HABITAT
Dry to moist woods and clearings; soils usually sandy.

DESCRIPTION
STEMS Horizontal stems on ground surface, forking, rooting at intervals, with annual constrictions. Erect branches at first simple, then branching; fertile branches with a leafy-bracted stalk bearing 1–5 sessile or short-stalked strobili (cones).

LEAVES Linear, tapered to translucent or white hairlike bristle, spreading to ascending and often incurved; margins entire or sometimes toothed.

STROBILI On long slender stalks with several branches, each branch tipped by a single strobilus. Bracts of strobili

coarsely fringed, at least the lower bracts with white hair-like tips.

SIMILAR SPECIES Identification of mature plants of running ground-pine is usually easy. The extended soft, hair-like bristles on the leaves help distinguish *Lycopodium* from other members of the family.

• **One-cone ground-pine** (*Lycopodium lagopus*) is distinguished, as the name implies, by having only a single strobilus per vertical stem, versus usually 2 or more in *L. clavatum*. Another useful distinction is the number of lateral branches from each vertical stem: *L. lagopus* generally has 2 or 3 branches, *L. clavatum* has 3–6 branches.

• Young or sterile plants may be confused with **interrupted clubmoss** (*Spinulum annotinum*), but the leaves in that species are tapered to a stiff bristle and not to a long hair-like tip (and when mature, the strobili are not stalked).

NAME This is the plant to which the term clubmoss was originally applied in Europe, and in 1753 Linnaeus accordingly gave it the species name *clavatum* (from Latin, *clava*, club, referring to its clublike stalked cones). It was later found in North America and at high elevations in the tropics.

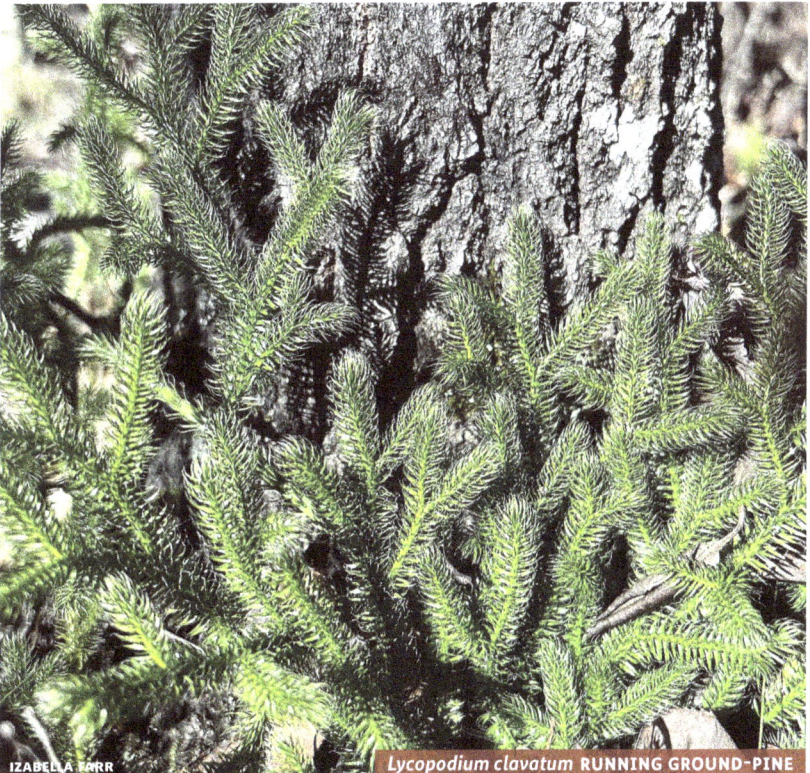

IZABELLA FARR

Lycopodium clavatum **RUNNING GROUND-PINE**

Lycopodium lagopus (Laestad. ex Hartm.) Zinserl. ex Kuzen.
ONE-CONE GROUND-PINE

DISTRIBUTION CT | MA | ME | NH | NJ | NY | PA | RI | VT

FIELD TIPS
- trailing, evergreen clubmoss
- stems densely covered with small upright leaves
- strobili (cones) usually single at end of stalk

HABITAT

Sterile, sandy acid soils in open mixed conifer-hardwood or coniferous forests, sandy borrow pits, fields, roadsides.

DESCRIPTION

STEMS Horizontal stems on or near ground surface, forking, rooting at intervals, with annual constrictions. Erect branches at first simple, then branching 1–2 times, the branches upright, with noticeable annual constrictions; fertile branches with a leafy-bracted stalk bearing 1–5 sessile or short-stalked strobili (cones).

LEAVES Linear, tapered to translucent or white hairlike bristle, ascending to appressed; margins entire or sometimes toothed.

STROBILI Single at end of each long, slender mostly unbranched stalk, (rarely with 2 strobili per stalk, these attached at same point). Sporophylls (bracts of strobili) tapered to hairlike tip.

SYNONYMS *Lycopodium clavatum* subsp. *megastachyon* (Fernald & Bissell) Selin, *Lycopodium clavatum* subsp. *monostachyon* (Hook. & Grev.) Seland., *Lycopodium clavatum* var. *brevispicatum* Peck, *Lycopodium clavatum* var. *integerrimum* Spring

SIMILAR SPECIES Similar to **running ground pine** (*Lycopodium clavatum*), but with leaves ordinarily ascending or appressed and strobilus single on a shorter peduncle.

NOTE In the past, this taxon, with one strobilus per stalk and more appressed leaves than in typical running ground-pine, has been considered a subspecies, variety, or form of *L. clavatum*.

NAME Lagopus from Greek, *lagos*, hare or rabbit, plus *pous*, foot, perhaps referring to the feathery appearance of the branches.

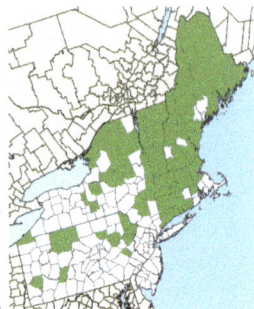

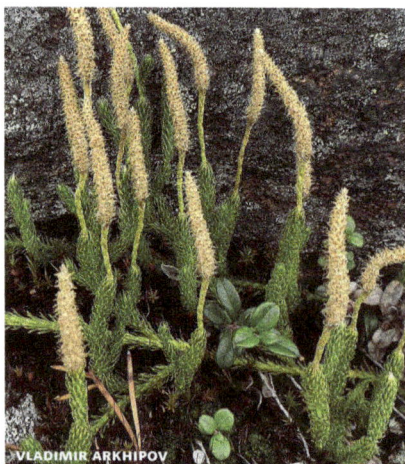

VLADIMIR ARKHIPOV

Spinulum A. Haines **BRISTLY CLUBMOSS**

EVERGREEN, TRAILING CLUBMOSS; upright stems prickly, with annual constrictions (resulting from shorter leaves being formed at end of one year's growth and beginning of the next year's growth). Sporangia borne in stalkless strobili at ends of branches. Worldwide, 3 species, all of which occur in North America; 2 species in our flora.

KEY CHARACTERS
- Plants with upright shoots, these branched near base.
- Branches round in cross-section.
- Leaves tipped with tiny spine.
- Strobili (cones) not stalked.

NAME From Latin, *spinula,* a small spine, as found at the leaf tips in *Spinulum.*

Spinulum annotinum

KEY TO *SPINULUM* | CLUB-MOSS

1 Longest leaves 6–9 mm long, spreading or reflexed, usually at least obscurely toothed near tip; regionwide . . . . *Spinulum annotinum*, INTERRUPTED CLUBMOSS, page 281
1 Longest leaves less than 6 mm long, all but lowest leaves ascending, usually entire; northern portions of region (ME, NH, NY, VT) . . . . *Spinulum canadense*
   . . . . NORTHERN CLUBMOSS, page 283

Spinulum annotinum (L.) A. Haines
INTERRUPTED CLUBMOSS

DISTRIBUTION CT | MA | ME | NH | NJ | NY | PA | RI | VT

FIELD TIPS
- evergreen trailing plants
- stems prickly to touch
- leaves spreading to reflexed
- strobili stalkless, single at stem ends

HABITAT
Wet to dry forests and clearings; soils often sandy or rocky.

DESCRIPTION

STEMS On ground surface or in uppermost humus layer, mostly unbranched, rooting at intervals, with annual constrictions. Erect stems simple or branched near base, 15–20 cm long, with pronounced annual constrictions.

LEAVES 8- to 10-ranked, more or less stiff and prickly, spreading to reflexed, widest above middle, tipped with a sharp point; margins finely toothed.

STROBILI Stalkless at ends of leafy stems; usually 1 strobilus per branch stem; sporophylls abruptly tapered to a sharp pointed tip.

SYNONYMS *Lycopodium annotinum* L., *Lycopodium dubium* Zoega

CONSERVATION STATUS *endangered* New Jersey, Rhode Island.

SIMILAR SPECIES **Shining fir-moss** (*Huperzia lucidula*) similar, but its leaves not spine tipped and it does not produce strobili.

NAME Linnaeus named this species *Lycopodium annotinum* from European specimens in 1753, but it was soon found in America as well. Epithet from Latin, *annotinus,* a year-old, perhaps referring to the annual constrictions 'interrupting' the stem.

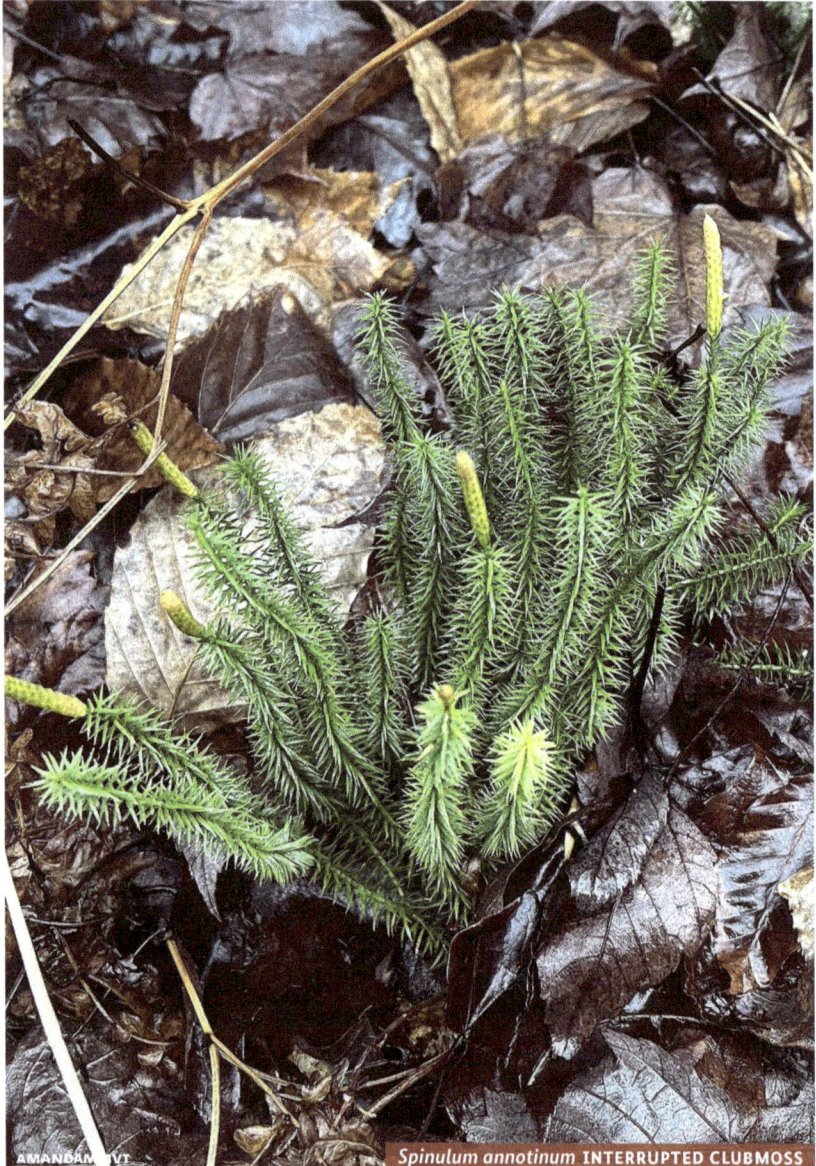

Spinulum annotinum **INTERRUPTED CLUBMOSS**

Spinulum canadense (Nessel) A. Haines
NORTHERN CLUBMOSS

DISTRIBUTION ME | NH | NY | VT

FIELD TIPS
- evergreen trailing plants
- stems prickly to touch
- leaves ascending to appressed
- strobili stalkless, single at stem ends

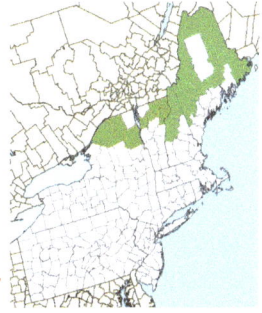

HABITAT

Open, alpine and subalpine ridges, rarely in cool, mountain peatlands.

DESCRIPTION

STEMS On ground surface or in uppermost humus layer, mostly unbranched, rooting at intervals, with annual constrictions. Erect stems simple or branched near base, 15–20 cm long, with pronounced annual constrictions.

LEAVES (of erect stems) 2.5–6 mm long, lance-shaped to lance-oblong, obscurely toothed, ascending to appressed against the stem; leaves just above annual constrictions broadest at or near base of leaf.

STROBILI Stalkless at ends of leafy stems; usually 1 strobilus per branch stem; sporophylls abruptly tapered to a sharp pointed tip.

SYNONYMS *Lycopodium annotinum* var. *pungens* (Bachelot de la Pylaie) Desvaux, *Lycopodium canadense* Nessel

SIMILAR SPECIES Interrupted clubmoss (*Spinulum annotinum*) similar, but leaves of *S. canadense* typically smaller, broadest at base and nearly entire; also reported to have usually more than 25 stomata on upper leaf surface. In *S. annotinum,* leaves usually larger, broadest near middle and stomata usually absent.

NOTE Perhaps best considered a minor variant of *Spinulum annotinum;* the two taxa are sometimes found growing together and are reported to hybridize.

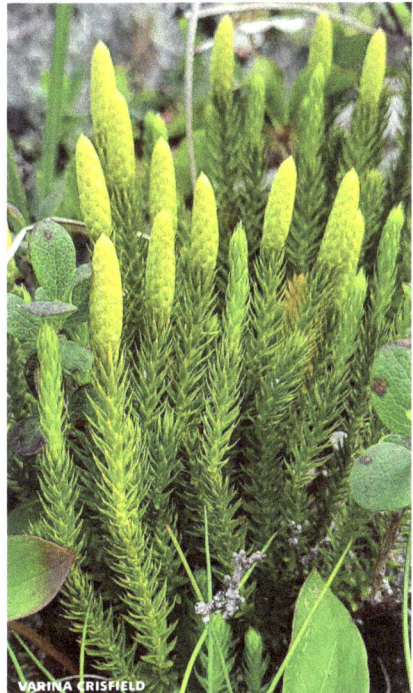

SELAGINELLACEAE *Spike-moss family*

Selaginella Beauv. SPIKE-MOSS

LOW, CREEPING MOSS-LIKE PLANTS with branching stems and few fine roots. Leaves simple, overlapping, 4-ranked or spirally arranged, with or without a bristle tip. Spores of two types, with some sporangia containing megaspores, others with microspores, borne in axils of leaf-like sporophylls of a terminal cone. Megaspores germinate into the female, egg-bearing gametophyte; the smaller microspores form the male, sperm-bearing gametophyte. Worldwide, the family contains five genera and somewhat more than 300 species, mainly in tropical and subtropical regions, and nearly all of which are in genus *Selaginella*. *Selaginella* is the sole genus, with about 700 species; four species in our flora.

KEY CHARACTERS
• Small, creeping plants.
• Leaves small and overlapping on stem.
• Spores of 2 types, borne in sporangia in axils of leaf-like sporophylls at ends of stems.

NAME Diminuitive of *selago*, an ancient name of a clubmoss (Lycopodiaceae), which *Selaginella* resemble.

KEY TO *SELAGINELLA* | SPIKE-MOSS

1   Sterile leaves in a spiral around stem, and all of similar length and width 2
1   Sterile leaves 4-ranked around stem, and of 2 sizes 3
2   Sterile leaves tipped with a short, rough bristle (this maybe absent in older leaves); strobili appearing 4-angled from the 4-ranks of appressed sporophylls (leavings bearing sporangia); plants of dry, sandy or rocky sites; regionwide . . . . *Selaginella rupestris*
. . . . SAND CLUBMOSS, page 287
2   Sterile leaves tapered to a sharp point but not bristle-tipped; strobili cylindric, the sporophylls widely spreading; plants of fens and calcareous shores; rare in ME . . . .
. . . . *Selaginella selaginoides*, NORTHERN SPIKE-MOSS, page 288
3   Sterile leaves of mid-portion of stem abruptly tapered tip, upper part of leaf underside keeled (like a boat hull); regionwide . *Selaginella apoda*, MEADOW SPIKE-MOSS, page 284
3   Sterile leaves of mid-portion of stem long-tapered to tip, not keeled; uncommon in CT, NY . . . . *Selaginella eclipes*, HIDDEN SPIKE-MOSS, page 286

Selaginella apoda (L.) Spring
MEADOW SPIKE-MOSS

DISTRIBUTION CT|MA|ME|NH|NJ|NY|PA|RI|VT
FIELD TIPS
• delicate, creeping, evergreen plant
• leaves thin and translucent, mosslike but with true veins
• stems threadlike, pale green
HABITAT
Wet woods, swamps and shores.

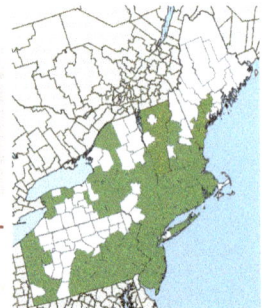

DESCRIPTION
STEMS Delicate, freely branching, forming small open mats.
STERILE LEAVES Thin and membranous, yellow-green, 4-ranked; of 2 sizes, the lateral 2 rows larger and spreading, the dorsal and ventral leaves smaller and appressed; with keeled midrib on upper portion of leaf underside, the midrib stopping short of leaf tip; margins very finely toothed.
STROBILI Somewhat 4-angled, at ends of branches.
SYNONYM *Selaginella apus* Spring
CONSERVATION STATUS *endangered* Maine.
SIMILAR SPECIES
• **Hidden spike-moss** (*Selaginella eclipes*) similar but median leaves in that species long tapered to a bristly tip. In *S. apoda,* leaves tapered to only a pointed, not bristly, tip.
NAME This plant was known to Linnaeus from "Carolina, Virginia, and Pennsylvania," and named by him in 1753 *Lycopodium apodum.* Upon its transfer to the genus *Selaginella* by Spring in 1840 the species name was wrongly spelled *apus,* as pointed out by Fernald in 1915.

Selaginella apoda MEADOW SPIKE-MOSS
PATRICIA BUTTER

Selaginella eclipes Buck
HIDDEN SPIKE-MOSS

DISTRIBUTION CT | MA *historical* | NY

FIELD TIPS
- creeping, forming small mats
- stems with threadlike branches
- leaves 4-ranked, of 2 sizes

HABITAT

Open areas of fens, wet sandy, rocky, or marly lake shores and river banks, seeps, openings in wet meadows, usually in strongly calcareous sites.

DESCRIPTION

STEMS Creeping, forming small mats; fertile branch tips only slightly elevated above ground surface.

STERILE LEAVES Of 2 types, the lateral leaves elliptic to oval, at right angles to stem; the other leaves smaller, tipped with long transparent bristle; leaves not keeled, the midrib extending to leaf tip; margins with very small teeth (visible under a hand lens).

STROBILI Loose, flattened, single or in pairs.

SYNONYM *Selaginella apoda* subsp. *eclipes* (W.R. Buck) Skoda

SIMILAR SPECIES **Meadow spike-moss** (*Selaginella apoda*) similar but median leaves tapered to only a pointed, not bristly, tip as in *S. eclipes*. The leaves of meadow spike-moss are also keeled on their underside near the tip.

NOTE Hidden spike-moss was not described as a distinct species until 1977, and because of many similarities, is often treated as a subspecies of *Selaginella apoda*.

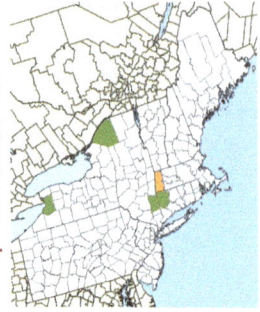

JIENNE

Selaginella eclipes HIDDEN SPIKE-MOSS

Selaginella rupestris (L.) Spring
SAND CLUBMOSS

DISTRIBUTION CT | MA | ME | NH | NJ | NY | PA | RI | VT

FIELD TIPS
- stiff, evergreen, mat-forming plant
- stems covered with appressed leaves
- strobili square-sided
- dry, rocky or sandy habitats

HABITAT

Dry rocky balds, sand dunes and open sandy areas; soils acidic; vegetation sparse, *Polytrichum* moss often present.

DESCRIPTION

STEMS Often forked at ends, covered with overlapping, appressed leaves; forming open, evergreen, gray-green mats.

STERILE LEAVES Narrowly lance-shaped, about 3 mm long, green or sometimes reddish, tipped with white or transparent bristle, grooved on the underside midrib; margins distinctly spiny.

STROBILI 4-sided, at ends of branches.

SYNONYM *Lycopodium rupestre* L.

NOTE Plants bright green when fresh and moist, turning gray-green when dry.

NAME In 1753 Linnaeus gave the name *Lycopodium rupestre* to a plant said to grow in "Virginia, Canada, and Siberia." In 1805 Beauvais proposed the genus *Selaginella* for members of the family having 2 sizes of spores, and in 1838, Spring placed Linnaeus' plant in the Spike-Moss Family.

Selaginella rupestris **SAND SPIKE-MOSS**

ULSTERBOTANY

Selaginella selaginoides (L.) Beauv. ex Mart. & Schrank
NORTHERN SPIKE-MOSS

DISTRIBUTION ME

FIELD TIPS
- mat-forming plant
- leaves with spine-like teeth
- strobili cylindric, on upright stems

HABITAT

Circumneutral fens and fen-like shores; in sun or partial shade; often on mossy hummocks.

DESCRIPTION

STEMS Creeping and branching, forming small mats. Fertile branches upright, with lower leaves similar to those of the stem but the leaves becoming larger upwards to form the sporophylls of the nearly cylindric strobilus.

STERILE leaves 2–4 mm long, spreading or ascending, tapered to pointed tip; margins coarsely toothed with soft, spine-like projections.

STROBILI Cylindrical at ends of upright stems.

CONSERVATION STATUS *threatened* Maine; a circumboreal species more common north of our region.

SIMILAR SPECIES Northern spike-moss similar in habit to **northern bog clubmoss** (*Lycopodiella inundata*) with its creeping sterile stems and erect fertile branches tipped by slightly expanded strobili; however, leaf margins of northern bog clubmoss smooth and not spiny-toothed.

NOTE Plants may be difficult to see as sometimes partially buried in moss or hidden by the surrounding taller vegetation.

MARTIN PRINZ

Selaginella selaginoides NORTHERN SPIKE-MOSS

REFERENCES

IDENTIFICATION GUIDEBOOKS

Cobb, B., C. Lowe and E. Farnsworth. 2005. *A field guide to ferns and their related families.* New York Houghton Mifflin Company.

Cody, W. J., and D. M. Britton. 1989. *Ferns and fern allies of Canada.* Canadian Government Publishing Center, Ottawa.

Cranfill, Ray. 1980. *Ferns and fern allies of Kentucky.* Kentucky Nature Preserves Commission.

Flora of North America Editorial Committee (eds). 1993. *Flora of North America, North of Mexico, Vol. 2.* Oxford University Press, New York.

Gilman, A.V. 2002. *Ophioglossaceae of Vermont.* V.F. Thomas Company, Bowdoin, Maine.

Gleason, H. A. and A. Cronquist. 1991. *Manual of vascular plants of northeastern United States and adjacent Canada.* 2nd edition. New York Botanical Garden, New York.

Hallowell, A. C., and B. G. Hallowell. 2001. *Fern finder: A guide to native ferns of central and northeastern United States and eastern Canada.* Nature Study Guild Publishers. Rochester, New York.

Haines, A. 2011. *Flora Novae Angliae: A manual for the identification of native and naturalized higher vascular plants of New England.* Yale University Press.

Harris, J. G. and M. W. Harris. 1994. *Plant identification terminology; an illustrated glossary.* Spring Lake Publishing. Spring Lake, Utah.

Lellinger, D. B. 1985. *A field manual of the ferns and fern-allies of the United States and Canada.* Smithsonian Institution Press. Washington, D.C.

Magee, D. W. and H. E. Ahles. 1999. *The flora of the northeast: A manual of the vascular flora of New England and adjacent New York.* The University Press of Massachusetts.

Mickel, J. T. 1979. *How to know the ferns and fern allies.* W. C. Brown Company Publishers. Dubuque, Iowa.

Odum, Eugene C. 1981. *Field guide to northeastern ferns.* New York State Mueum Bulletin 444. Albany, New York.

Parsons, F. T. 1961. *How to know the ferns: A guide to the names, haunts, and habits of our common ferns.* Dover Publications, New York.

Shaver, J. M. 1970. *Ferns of the eastern central states, with special reference to Tennessee.* Dover Publications, New York.

Snyder, L. H., Jr., and J. G. Bruce. 1986. *Field guide to the ferns and other pteridophytes of Georgia.* University of Georgia Press, Athens.

Thomas, G. B. and S. W. Francis. 2004. *Wildflowers and ferns of Kentucky.* University Press of Kentucky.

Tyron, R. M. and R. C. Moran. 1997. *The ferns and allied plants of New England.* Massachusetts Audubon Society.

Wherry, E. T. 1972. *The southern fern guide.* Doubleday, Garden City, New York.

Wherry, E. T. 1961. *The fern guide; northeastern and midland United States and adjacent Canada.* Doubleday. New York.

FERN CULTURE

Foster, F. G. 1984. *Ferns to know and grow.* 3rd ed. Timber Press.

Hoshizaki, B. J. and K. A. Wilson. 1999. *The cultivated species of the fern genus Dryopteris in the United States.* American Fern Journal 89: 1–100.

Hoshizaki, B. J. and R. C. Moran. 2001. *Fern grower's manual.* Timber Press.

Jones, D. L. 1987. *Encyclopaedia of ferns.* Timber Press.

REFERENCES

Mickel, J. T. 1994. *Ferns for American gardens*. MacMillan Publishing Company, New York.

Moran, R. C. 2004. *A natural history of ferns*. Timber Press.

Olsen, Sue. 2007. *Encyclopedia of garden ferns*. Timber Press.

SELECTED TECHNICAL REPORTS

Beitel, J. M. 1979. *The clubmosses Lycopodium sitchense and L. sabinifolium in the Upper Great Lakes Region*. Michigan Botanist 18:3–13.

Blasdell, R.F. 1963. *A monographic study of the fern genus Cystopteris*. Memoirs of the Torrey Botanical Club Vol. 21, No. 4.

Chase M. W. and J. L. Reveal. 2009. *A phylogenetic classification of the land plants to accompany APG III*. Botanical Journal of the Linnean Society 161: 122–127.

Christenhusz, M. J. M., X.-C. Zhang, and H. Schneider. 2011. *A linear sequence of extant families and genera of lycophytes and ferns*. Phytotaxa 19 7–54.

Crane, E. H. 1997. *A revised circumscription of the genera of the fern family Vittariaceae*. Systematic Botany 32: 509–517.

Cranfill, R. 1983. *The distribution of Woodwardia areolata*. American Fern Journal 73 46–52.

Evrard, C. and C. Van Hove. *Taxonomy of the American Azolla Species (Azollaceae): A Critical Review*. Systematics and Geography of Plants Vol. 74, No. 2 (2004), pp. 301-318.

Farrar, Donald R. 2006. *Systematics of Moonworts (Botrychium subgenus Botrychium)*. Department of Ecology, Evolution and Organismal Biology, Iowa State University, Ames, Iowa.

Hickey, R. J. 1977. *The Lycopodium obscurum complex in North America*. American Fern Journal 67:45–49.

Johnson, D. M. 1986. *Systematics of the New World species of Marsilea (Marsileaceae)*. Systematic Botany Monographs 11: 1–87.

Koenemann DM, Maisonpierre JA, Barrington DS. 2011. *Broad-scale integrity and local divergence in the fiddlehead fern, Matteuccia struthiopteris (L.) Todaro (Onocleaceae)*. American Fern Journal 101: 213–230.

Kramer, K. U., and P. S. Green, eds. 1990. *Pteridophytes and Gymnosperms*. Vol. 1 of *The Families and Genera of Vascular Plants*, edited by K. Kubitzki. Springer-Verlag, Berlin.

Lehtonen, S. 2011. *Towards resolving the complete fern tree of life*. PLoS ONE 6(10): e24851.

Lumpkin, T. A., and D. L. Plucknett. 1980. *Azolla: Botany, physiology, and use as a green manure*. Economic Botany 34: 111–153.

Moore, A. W. 1969. *Azolla: Biology and agronomic significance*. The Botanical Review 35: 17–35.

Moran, R. C. 1982. *The Asplenium trichomanes complex in the United States and adjacent Canada*. American Fern Journal 72: 5–11.

Patel, Nikisha R., Susan Fawcett, and Arthur V. Gilman. *Phegopteris excelsior (Thelypteridaceae): A new species of North American tetraploid beech fern*. Novon: A Journal for Botanical Nomenclature 27(4), 211-218, (4 November 2019).

Pippen, R. W. 1966. *Lygodium palmatum, the climbing fern, in Michigan*. Michigan Botanist 5: 64–66.

PPG-I. 2016. *A community-derived classification for extant lycophytes and ferns*. Journal of Systematics and Evolution, vol. 54, no. 6, pp. 563-603.

Rothfels, C. J., M. A. Sundue, Li-Yaung Kuo, A. Larsson, Masahiro Kato, E. Schuettpelz, and K. M. Pryer. *A revised family-level classification for eupolypod II ferns (Polypodiidae Polypodiales)*. Taxon 61 (3) June 2012 515–533.

Smith, A.R., K.M. Pryer, E. Schuettpelz, P. Korall, H. Schneider, and P. G. Wolf. 2006. *A classification for extant ferns*. Taxon 55 705–731.

Stevens, P. F. 2001 onwards. Angiosperm Phylogeny Website. Version 12, July 2012 (and continuously updated since). www.mobot.org/MOBOT/research/APweb/.

Tyron, R. M. and A. F. Tyron. 1982. *Ferns and allied plants with special reference to tropical America*. New York Springer-Verlag.

Wagner, G. M. 1997. *Azolla: A review of its biology and utilization*. Botanical Review 63: 1–26.

Wagner, W.H. Jr., and F.S. Wagner. 1990. *Moonworts (Botrychium subg. Botrychium) of the upper Great Lakes region, U.S and Canada, with descriptions of two new species*. Contr. Univ. Mich. Herb. 17:313-325.

Wagner, W.H. Jr. and F.S. Wagner. 1981. *New species of moonworts, Botrychium subg. Botrychium (Ophioglossaceae), from North America*. Amer. Fern J. 71:20-30.

FERN LEAF DIVISIONS, STIPE BUNDLES, SORI

pinnatifid

pinnate

pinnate-pinnatifid

bipinnate

bipinnate-pinnatifid

tripinnate

tripinnate-pinnatifid

stipe with 1 small bundle

stipe with 1 horshoe-shaped bundle

stipe with 2 circular bundles

stipe with 2 linear bundles

stipe with 3-9 all circular bundles

stipe with 3-many, some linear, bundles

sori circular and marginal

sori circular not marginal

sori linear and marginal

sori linear not marginal

fronds dimorphic

GLOSSARY

See Introduction (page 12) for an illustrated glossary of basic fern terms.

abaxial of the side or surface of an organ, facing away from the axis, e.g. the lower or dorsal surface of the lamina. Compare to adaxial.

acicular stiff and needle-like.

acroscopic pointing towards the apex. Compare to basiscopic.

acrostichoid with sporangia densely covering the abaxial surface of the lamina, as in the tropical fern *Acrostichum*.

acuminate gradually tapering to a protracted point.

acute terminating in a sharp or well-defined point.

adaxial of the side or surface of an organ, facing towards the axis, e.g. the upper or ventral surface of the lamina. Compare to abaxial.

adnate fused to an organ of a different kind, as a pinna to a rachis.

alate winged.

alete used of a spore which forms alone, i.e. not in diads or tetrads.

allotetraploid having 4 complete sets of homologous chromosomes resulting from interspecies hybridization followed by chromosome doubling; a plant with such a condition.

anadromous a type of venation in which the first set of veins in each segment of the frond arising from the side of the midrib facing toward the frond tip.

anastomosed veins united to form areoles; net-veined.

annular arranged in or forming a ring.

annulus the elastic ring of cells in a sporangium that initiates dehiscence; a complete or partial ring or cluster of thick-walled cells on the spore case functioning to open the spore case.

antheridium (plural antheridia) the male sex organ containing the sperm; borne on the underside of the prothallus.

antrorse bent, and pointing towards the apex. Compare to retrorse.

apex (plural apices) the tip, the point farthest from the point of attachment.

apiculus a small abrupt flexible point at the apex of a pinna or pinnule. adj. apiculate.

apogamous applied to ferns in which a sporophyte develops from gametophyte cells other than a fertilized egg.

apogamy a form of asexual reproduction in which new sporophytes are produced directly from the prothallus tissue instead of from the fertilized egg cell (zygote).

apomict a plant that produces viable spores without fertilization.

apospory a form of asexual reproduction in which prothalli are produced directly from young sporophyte tissue instead of from spores.

appressed pressed closely against a surface (or another organ) but not united with it.

arachnoid composed of fine tangled hairs like a cobweb.

archegonium the female sex organ containing the egg; the structure that produces the female gamete or egg. plural archegonia.

areolate having netted veins.

areole a space enclosed by netlike veins (also see reticulum).

aristate having a stiff bristle-like tip.

articulate jointed; having joints where separation may occur naturally.

attenuate pinnae gradually tapering to a very narrow, slender point.

auricle an ear-like lobe at the base of a lamina, pinna or pinnule (adj. auriculate).

axes collective term referring to the petiole, rachis, costae, costules, and midrib of a frond.

basal pinnae (basal pinna pair) the lowermost pair of pinnae, closest to the common stalk.

basiscopic pointing towards the base; the basiscopic margin of a pinna is that which faces the base of the leaf; the basiscopic margin of a pinnule is that which faces the base of the pinna. Compare to acroscopic.

bicolorous having two colors, usually referring to scales in which the central part is darker than the margins. Compare to concolorous.

binomial the species name, the two words consisting of the genus name and the specific epithet.

bipinnate twice pinnately branched (2-pinnate).

bipinnatifid twice pinnatifid; the divisions of a pinnatifid frond are again pinnatifid (2-pinnatifid).

blade the flat, expanded portion of a leaf.

bristle a stiff hair which is more than one cell broad at the base.

bulbiferous bearing bulbils.

bulbil or bulblet small bulb-like body borne upon a stem or leaf and serving to vegetatively reproduce the plant.

catadromous a type of venation in which the first basal branch or vein (as on a pinna) arises from the side toward the frond base. Compare to anadromic.

caudate with a long and thin tip.

caulescent developing an aerial stem or trunk.

cell the basic unit of plant structure consisting, at least when young, of a protoplast surrounded by a wall.

chartaceous thin and papery.

chlorophyll pigment(s) constituting the green coloring matter of plants and absorbing radiant energy in photosynthesis.

ciliate fringed with hairs.

circinate (circinate vernation) coiled in a spiral with the tip innermost, as in the fiddleheads of many ferns.

clathrate latticed or pierced with apertures like a trellis, as in the rhizome scale of *Asplenium* (a hand lens or microscope is needed to see this).

clavate club-shaped.

cleft technically, deeply cut, usually at least half-way to the middle or base.

commissure a juncture or seam; in *Pellaea,* a ±continuous marginal sorus formed when laterally expanded fertile vein endings coalesce.

common stalk the stalk below the junction of the sporophore and trophophore.

compound of a leaf, having the blade divided into 2 or more distinct leaflets.

concolorous colored uniformly; the same color on both sides.

conduplicate folded flat together lengthwise; of developing leaves.

cone tight cluster of highly modified, spore-bearing leaves borne at branch tips.

confluent flowing or running together.

convergent tending to come together, merging.

cordate of a leaf blade, broad and notched at the base; heart-shaped (in 2 dimensions).

coriaceous leathery.

corm in Isoetes, the condensed stem, which may be 2-5-lobed.

costa (plural costae) the midrib of a pinna.

costule the midrib of a pinnule or segment of lower order, except the central vein of an ultimate segment which is usually termed the midrib.

crenate with small, rounded teeth; scalloped.

crenulate with very small rounded teeth along the margin.

crested having forked tips, usually many; usually refers to the frond, pinnae, or segments.

cristate in ferns, having a tasselled margin to the fronds.

crown the tip of the stem where the leaves arise; usually applied to thick, upright stems.

crozier uncurling frond; a young coiled fern frond; the fiddlehead.

cuneate wedge-shaped, with the narrow part at the point of attachment, usually referring to the base of a pinna or pinnule.

cuspidate tipped with a cusp or a sharp and firm point.

cut (cutting) referring to the presence of dissection or division.

deciduous shed seasonally, not evergreen..

decurrent referring to the bases of leaves or pinnae continuing down the stem or rachis beyond the point of attachment.

decussate borne in pairs alternately at right angles to each other.

deflexed bent abruptly downwards.

deltate or deltoid triangular with the sides of about equal length.

dentate of margins, toothed; usually with broad teeth directed outward.

denticulate finely toothed.

dichotomous forked regularly into pairs.

dimorphic having 2 different forms; in ferns the term usually refers to differences in size or shape of the sterile and fertile fronds (or segments). Compare to monomorphic.

dioecious having the male and female reproductive structures on separate plants. Compare to monoecious.

diploid having 2 of the basic sets of chromosomes in the nucleus; a plant with such a condition. Compare to haploid, polyploid.

dissected cut into lobes or lobe-like segments.

distal furthest point; remote from the point of origin or attachment. Compare to proximal.

divergent spreading apart.

divided cut to the base or midrib into lobes or segments. As applied to ferns, "divided" is a generic term meaning "dissected" and includes pinnate (cut to midrib) and pinnatifid (cut into lobes without divisions reaching the midrib).

dorsal relating to the back or lower side of a leaf.

echinate spiny, referring to a spore.

eglandular without glands.

elaters in *Equisetum,* appendages of the spore which help in dispersal; elators are formed from the outermost wall layer, coiling and uncoiling as air is dry or moist.

elliptic ellipse shaped, broadest in the center and narrower at the two equal ends.

elongate drawn out, lengthened.

emarginate having a shallow notch in the margin.

emersed standing out or rising above a surface (as of water).

entire undivided; the margin continuous, not incised, lobed or toothed.

ephemeral lasting only a short time; (of indusium: quickly shed).

epilithic growing on rocks; also epipetric, saxicolous, or rupestral.

epiphyte (epiphytic) a plant that grows upon another plant; an epiphyte uses its host plant only for support; it is not parasitic.

erose of a margin, finely and irregularly eroded or incised.

euphyllophytes the group containing both ferns (monilophytes) and seed-bearing plants (spermatophytes).

eusporangiate having sporangia with walls more than one cell thick. Compare to leptosporangiate.

excurrent vein a vein running toward the margin, not the midrib.

exindusiate lacking an indusium.

extirpate to destroy completely; eradicate, in conservation usually refers to a species historically present but now absent from a state.

falcate sickle-shaped, curved and flat.

false indusium an indusium formed by the rolled-over margin of the leaf enclosing the sorus, as in many genera in the *Pteridaceae, such as Adiantum, Cheilanthes* and *Pellaea.*

false veins rows of thickened cells in a leaf/leaflet which are not part of the vascular system, as in some genera of the Hymenophyllaceae.

family taxonomical division comprised of a group of related genera.

farina a meal-like powder, usually white or yellow and found on the lower surfaces of fronds; characteristic of fern genera such as *Argyrochosma, Pentagramma,* and *Pityrogramma* (these not present in our region).

farinose covered with a meal-like powder (farina).

fern allies group of vascular plants traditionally thought to be closely related to ferns, based mainly on the spore-bearing characteristic. These comprise the families of horsetails (Equisetaceae), quillworts (Isoetaceae), clubmosses (Lycopodiaceae), whisk ferns (Psilotaceae) and spikemosses (Selaginellaceae). Recent research has shown quillworts, clubmosses and spikemosses to be much less closely related to ferns, while horsetails and whisk ferns are more closely related, in fact essentially part of the main fern grouping (reflected in the organization of this flora).

fertile in ferns usually referring to leaves that bear sori.

fiddlehead a fern leaf, young or in bud, that is coiled in a spiral pattern. Also see crozier.

filiform threadlike.

fimbriate of a margin, fringed with fine hairs.

flabellate fan-shaped.

floccose covered with soft tangled woolly hairs.

flora list of all the species growing in a region, or a collective term for all the species growing in a region; also refers to books that identify the plants within a certain geographical area.

frond the whole leaf of a fern, including the blade and stipe (petiole).

fruit-dots sori, clusters of sporangia.

fugacious shed or withering away very early.

fused grown to or united with a similar part.

gametophyte a small, usually flat plant, bearing the sex organs (archegonia and antheridia) that in turn produce the gametes. Each cell in the body of the gametophyte has one set of chromosomes (1n); gametophytes grow from spores.

gene a unit on a chromosome that determines the inheritance of a particular trait.

gemma (plural gemmae) a bud or bulbil that detaches from the main plant and develops into a new plant, as in Huperzia selago.

gemmiferous bearing gemmae (asexual buds or bulbils).

genus (plural genera) term for describing a closely related group of species.

glabrescent becoming glabrous (smooth).

glabrous without hairs or scales; smooth.

gland a structure with a secretory function, embedded or projecting from the surface of the plant.

glandular having glands or functioning as a gland.

glaucous dull green with a bluish white or white, usually waxy, lustre.

globose almost spherical.

haploid with one set of chromosomes in the nucleus. Compare to diploid, polyploid.

hastate spear-shaped; of a leaf blade, narrow and pointed but with two basal lobes spreading approximately at right angles.

herbaceous soft in texture; midway in thickness between membranous and coriaceous, usually applied to the leaf; without a persistent woody stem above ground; dying back to the ground at the end of the growing season; leaf-like in color and texture.

heterophyllous having leaves that are not uniform along a branch, e.g. in Huperzia, with long leaves in the lower portions and smaller reduced leaves distally. Compare to homophyllous.

heterosporous having 2 kinds of spores (male and female, or microspores and megaspores). Compare to homosporous.

hirsute bearing coarse rough relatively long hairs. Compare to villous.

homophyllous with all leaves uniform along a branch. Compare to heterophyllous.

homosporous producing only one type of spore from which develops a gametophyte producing both male and female gametes. Compare to heterosporous.

hyaline translucent, almost like clear glass.

hybrid result of a sexual cross between 2 different taxa.

hydathode an enlarged vein tip on the upper surface of a blade; it often secretes water and minerals; in some ferns, the minerals may accumulate as a white deposit over the hydathode.

imbricate overlapping, like roof tiles.

incised cut deeply, sharply and often irregularly (an intermediate condition between toothed and lobed).

incurved bent or curved inwards or upwards; of leaf margins, curved towards the adaxial surface.

indusiate bearing an indusium.

indusium (plural indusia) the covering of a sorus, either a specialised organ or the incurved margin of the blade (also see false indusium).

internode the portion of a stem or other structure between two nodes.

introduced a plant that was brought into the country, region, or state (either deliberately or accidentally) by humans.

involucre the indusium of members of the Hymenophyllaceae.

jointed able to separate naturally at a certain point, leaving a scar; articulate.

laciniate cut into narrow-pointed lobes.

lacuna (plural lacunae) of Isoetes, a cavity within the leaves.

GLOSSARY

lamina (plural laminae) the 'blade' of a frond.

lanceolate lance-shaped; long and narrow, but broadest at base and gradually tapering to apex.

lateral situated on or arising from the side of an organ.

leaf frond used here to include both the 'leafy part' and the stipe or stem.

leaf blade the 'leafy' part of the frond excluding the stipe or stem.

leaflet one of the divisions in a compound leaf.

leaf-segment any subdivision of a frond. See pinna, pinnule, pinnulet.

leptosporangiate having sporangia with the walls only one cell thick. Compare to eusporangiate.

ligulate bearing a ligule; strap-shaped.

ligule a membranous structure towards the base of the upper leaf surface; in Isoetes a small triangular or elongate delicate tissue extending slightly above the sporangium.

linear long and narrow, much longer than wide, the sides parallel or nearly so.

lithophytic growing on rock; also see epilithic.

lunate shaped like a crescent or half-moon.

lobe a division or segment of an organ; technically, cut not over half-way to the middle or base, the sinuses and apex of segments rounded. Compare to cleft.

lobed bearing lobes.

lustrous shiny or glistening, glossy.

lycophytes the group comprising quillworts, clubmosses and spikemosses. Distinct from both euphyllophytes, the group that contains both seed plants (spermatophytes) and ferns (monilophytes).

macrospore the larger kind of spore in Selaginellaceae and Isoetes, and in other genera; also called megaspore.

marcescent withering without falling off; in contrast to jointed or articulate, which indicate withering and falling off at a joint.

massula (plural massulae) group of microspores enclosed in a hardened mucilage.

margin (marginal) the edge, as in the edge of a leaf blade.

medial being or occurring in the middle.

megasporangium the larger of the two kinds of sporangia produced in the sexual life cycle

of a heterosporous plant; produces megaspores.

megaspore the larger of the two kinds of spores produced in the sexual life cycle of a heterosporous plant, giving rise to the female gametophyte. They may be monomorphic as in Selaginella, or polymorphic as in some Isoetes species. Compare to microspore.

megasporocarp a sporocarp containing megasporangia.

megasporophyll a specialised leaf upon which (or in the axil of which) one or more megasporangia are borne.

meiosis the type of cell division that gives rise to spores; during meiosis, the cell replicates its chromosomes once and divides twice; the result is four cells with only half the chromosome number of the original cell.

micron one thousandth of a millimeter.

microsporangium the smaller of the 2 kinds of sporangia produced in the sexual life cycle of a heterosporous plant; as in Selaginella and Isoetes.

microspore the smaller of the 2 kinds of spores produced in the sexual life cycle of a heterosporous plant, giving rise to the male gametophyte. Compare to megaspore.

microsporocarp a sporocarp containing microsporangia.

midrib the central, and usually the most prominent, vein of a leaf or leaflike organ.

monoecious having the male and female reproductive parts in separate organs but on the same plant. Compare to dioecious.

monolete of a spore, bilateral, having a single straight scar.

monomorphic of uniform shape and size. Compare to dimorphic.

monilophytes the group comprising all ferns, now known to include horsetails and whiskferns; a recently-coined term to distinguish them from spermatophytes (seed-bearing plants).

midrib the central rib or vein of a leaf or other organ.

mucro a sharp abrupt terminal point. adj. mucronate.

mycorrhiza an association of a fungus with the root of a higher plant.

native occurring naturally, not introduced by humans.

naturalized an introduced plant that has become established and propogates itself naturally.

node the point on a stem where a leaf emerges.

nodosity in *Adiantum,* a callus or swollen node, often lacking normal coloration, where a pinna or pinnule stalk arises from a rachis.

nothosubspecies (often abbreviated to nothosubsp., nothossp. or n-subsp) hybrid where one or both parents is a subspecies (the term nothospecies is occasionally used to also denote a hybrid between two species).

ob- a prefix signifying the opposite of.

oblanceolate lanceolate with the broadest part near the tip.

obovate ovate or egg-shaped with the broader end at tip.

oblong 2 to 4 times longer than wide and the sides parallel or nearly so.

obtuse blunt or rounded at the tip.

orbicular essentially circular.

ovate egg-shaped in outline and broadest at base.

pallid pale.

palmate having veins, lobes or segments which radiate from a single point, as in maple leaves.

palmatifid of a leaf, deeply (but not completely) divided into several lobes which arise (almost) at the same level.

paraphysis (plural paraphyses) a sterile hair occurring among the sporangia of some ferns.

pectinate comblike.

pedate of a palmate or palmately-lobed leaf, having the lateral segments divided again.

pedicel the stalk of a sporangium.

peduncle the stalk of a sporocarp, e.g. in *Marsilea.*

peltate having the stalk attached to the lower surface usually at or near the center; like the handle on an open umbrella.

pendulous hanging or drooping.

perispore the folded membrane of most spores, forming an ornamental external covering.

persistent remaining attached to the plant beyond the expected time of falling.

petiole a leaf stalk, also known as a stipe.

phyllopodium (plural phyllopodia) a stumplike extension from the rhizome to which the fronds are attached, usually by a distinct abscission layer.

pilose hairy, the hairs soft and clearly separated but not sparse.

pinna (plural pinnae) the primary division of a pinnately divided frond; a leaflet; pinnae may be further divided into pinnules and again into pinulules.

pinna segment division of a pinna, whether cut entirely to the midrib or not.

pinna span the width of a fan-shaped pinna at its distal (outer) margin, measured in degrees of a circle, estimated by imagining a circle with the point of pinna attachment at its center; used most commonly with some species of *Botrychium.*

pinnate having a feather-like arrangement, divided into pinnae, with the pinnae (leaflets) arising at points along the rachis.

pinnate-pinnatifid referring to a blade that is once-pinnate and with the pinnae deeply lobed or cut, but not to their midrib.

pinnatifid cut half to three-fourths to the rachis.

pinnatisect cut almost all the way to the rachis but having the segments confluent with it.

pinnule a division of a pinna (i.e., a pinna lobe); the secondary pinna.

pinnulets divisions of pinnules.

polymorphic having more than 2 distinct morphological variants.

polyploid having more than 2 of the basic sets of chromosomes in the nucleus. Compare to diploid, haploid.

procumbent lying flat along the ground.

proliferous having adventitious leaf buds which produce new plants.

prostrate lying flat upon the ground.

prothallus (plural prothalli) the gametophyte stage of the fern; this is the independent stage where sexual reproduction takes place; in most ferns, it is a small, flattish, often roughly heart-shaped body.

protostele a simple primitive type of stele having a solid central vascular core.

proximal closest, usually referring to the part of leaf or leaf segment closest to the point of attachment.near to the point or origin of attachment. Compare to distal.

pseudo- false; apparent but not genuine.

pteridophyte traditional term encompassing both ferns and "fern allies"; the latter now known to be composed of some groups which are essentially ferns but look unlike them, and others which are not at all closely related. See Lycophytes, Monilophytes, Euphyllopytes.

puberulent minutely hairy.

pubescent clothed with short soft erect hairs.

rachis the midrib of a compound frond; i.e., the main axis above the lowermost primary pinna; the upper part of the petiole, bearing the pinnae and continuous with the stipe. Also spelled rhachis; plural rachises or rachides.

radial applied to a rootstock in which the fronds radiate and the roots are borne on all sides of the organ.

receptacle of ferns, the axis bearing the sporangia and sometimes also paraphyses. the tissue upon which the sporangia are borne. The receptacle is bristle-like in *Hymenophyllum* and *Trichomanes;* in most other ferns it is flush with the leaf surface or slightly elevated.

recurved curved or curled downwards or backwards.

reflexed bent sharply downwards or backwards.

reniform kidney-shaped.

reticulate (usually of veins): forming a network; net-veined.

reticulum a network, e.g. of veins. adj. reticulate.

retrorse bent, and pointing away from the apex. Compare to antrorse.

revolute rolled backward from the margins or apex. In ferns the term usually refers to a leaf margin rolled back to protect the sori (false indusium).

rhizoid a thread-like unicellular absorbing structure occurring, in the vascular plants, in gametophytes of ferns and some related plants; also present in many mosses which lack vascular systems.

rhizome the creeping (often underground) or climbing stem of a fern.

rhizophore in *Selaginella,* a leafless stem that produces roots.

rhomboidal diamond-shaped or almost so.

rootstock a short, erect stem.

rupestral growing on rocks; epilithic.

rugose deeply wrinkled.

rugulose covered with minute wrinkles.

saxicolous growing on rocks; epilithic.

scale small, often semi-transparent outgrowth of the outer layer of cells (epidermis), at least 2 cells wide to qualify as a scale, if only one cell wide considered a hair (trichome).

scalloped a series of semi-circles or curves resembling the edge of a scallop; with such a margin.

scandent climbing.

scarious thin, dry-looking, translucent, often whitish.

segment a division or part of a pinna.

septate divided internally by septa.

septum (plural septa) a partition.

sericeous clothed with silky hairs.

serrate having sharp, forward-pointing teeth; like a sawblade.

sessile without a stalk.

seta (plural setae) a stiff hair or bristle; in ferns these tend to be brown or black.

setose covered with bristles.

siliceous composed of or abounding in silica, as in some *Equisetum.*

simple undivided, not compound; of a frond, not divided into leaflets; of a hair or an inflorescence, not branched.

sinus membrane the membrane of a depression between adjacent lobes in a pinna, as in the Thelypteridaceae; the gap or indentation between teeth or lobes of frond.

soral flap the specialized fertile lobe in *Adiantum.*

sorus (plural sori) a cluster of spore cases (sporangia).

spathulate spoon-shaped; broad at the tip and narrowed towards the base.

species taxonomic division generally used to describe those plants that will interbreed freely with each other and share a range of visual similarities; the word is used for both the singular and plural.

spermatophytes seed-bearing plants.

spinulose bearing small spines over the surface.

sporangium (plural **sporangia**) a structure within which spores are formed; the globular organ in which the spores are produced.

spore a unicellular or few-celled sexual or asexual reproductive cell that germinates into a prothallus, which in turn gives rise to sexual reproduction.

sporeling tiny fern plant still attached to the gametophyte from which it has developed.

sporocarp a fruiting body containing sporangia; a round structure that contains sporangia within. Characteristic of 2 fern families: Salviniaceae and Marsileaceae; sporocarps are globose and delicate in the Salviniaceae, beanlike and hard in the Marsileaceae.

sporophore the fertile, spore-bearing portion of the frond.

sporophyll a specialised leaflike organ that bears one or more sporangia.

sporophyte the familiar plant that bears roots, stems, and leaves (as opposed to a gametophyte or prothallus); so-called because it is the phase of the life cycle that produces spores; each cell in the body of a sporophyte has two sets of chromosomes (i.e., is 2n).

sporulation the formation of spores.

stalk an unexpanded, unbranched supporting structure.

stele the vascular system of rhizome or stem, together with leaf traces.

stellate star-shaped.

sterile refers to leaves that do not produce sori and to hybrids in which spores are aborted.

stipe the stalk or petiole of the frond; that portion of the midrib of the frond between the rhizome and the lowermost primary pinna; does not bear pinnae.

stipule a basal appendage of a stipe or petiole, usually paired; in ferns, the term is sometimes applied to the flared leaf bases in *Osmunda* and *Osmundastrum*.

stolon runner from the main stem, producing a new plant that roots independently.

stomium the region of a sporangium at which dehiscence occurs and the spores are released.

stramineous straw-colored.

strigose having curved, sharp, forward-pointing hairs.

strobile an inflorescence resembling a spruce or fir cone, partly made up of overlapping bracts or scales.

strobilus a conelike body, as in the Lycopodiaceae and Selaginellaceae, consisting of sporophylls borne close together on the axis. plural strobili.

submersed growing, or adapted to growing, under water.

subspecies (often abbreviated to spp. or subsp.) subdivision of a species.

succulent juicy; fleshy.

sympatric with areas of distribution that coincide or overlap.

synonym an alternative scientific name, but not the one currently accepted by taxonomists.

taxon (plural **taxa**) any members of a specific taxonomic grouping, e.g., species, genus, etc.; one may refer to the *Asplenium trichomanes* and *Asplenium ruta-muraria* taxa (species); *Dryopteris* and *Polystichum* taxa (genus); or the Lycopodiaceae and Selaginellaceae taxa (family).

terete circular or almost so in cross section.

terminal at the tip.

ternate compounded into more or less equal divisions or groups of three.

terrestrial growing on the ground, not in trees or on rocks.

tetraploid having 4 of the basic sets of chromosomes in a nucleus; with 4n chromosomes per cell; a plant with such a condition.

tomentum a dense covering of woolly hair.

toothed with small lobes or teeth along the margin.

trichome an epidermal outgrowth, e.g. a hair (branched or unbranched), a papilla.

trigonous three-angled.

tripinnate pinnate with the pinnae and pinnules also pinnate (also 3-pinnate, thrice-pinnate).

triploid having 3 of the basic sets of chromosomes in the nucleus.

trophophore the sterile, foliaceous portion of the frond.

truncate ending abruptly, as if cut off.

ultimate segment the final, smallest divisions of a frond.

undulate gently wavy; with such a margin.

vallecula applied to the grooves in the intervals between the ridges, as in the stems of *Equisetum;* vallecular: pertaining to such grooves.

variety (often abbreviated to var.) subdivision of species, but less well-defined than a subspecies.

vascular pertaining to specialized tissue (xylem and phloem) that conducts water, mineral nutrients, and sugars.

vascular bundle the primary fluid-conducting system of a plant.

vein a strand of vascular tissue.

velum a membranous flap-like envelope which partially or wholly covers the sporangium, as in *Isoetes.*

venation the pattern formed by the veins in the leaf.

ventral belonging to the anterior or inner face of an organ, as opposed to dorsal; adaxial.

vernation the arrangement of the unexpanded fronds in a bud.

verticillate arranged in a whorl.

villous covered densely with fine long hairs but not matted. Compare to hirsute.

whorl leaves or other plant parts arranged in a circle around the stem.

wing a thin expansion or flat extension of an organ or structure.

xeric characterized by, relating to, or requiring only a small amount of moisture.

xerophyte a plant adapted to dry habitats.

xylem the tissue, in a vascular plant, that conducts water and mineral salts from the roots to the leaves.

zygote a fertilized egg cell; the first cell in the development of a sporophyte.

The Fern Gatherer, by Charles Sillem Lidderdale, 1877; painted during the fern craze that swept England in the second half of the 19th century.

INDEX

NOTE: Synonyms are listed in *italics.*

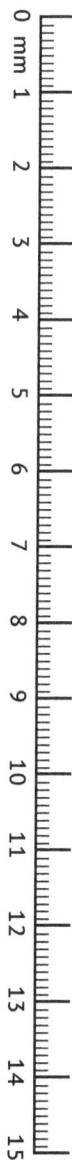

www.ingramcontent.com/pod-product-compliance
Lightning Source LLC
Chambersburg PA
CBHW052109030426
42335CB00025B/2901